T0256637

The Doctor's Garden

The Doctor's Garden

Medicine, Science, and Horticulture in Britain

Clare Hickman

Yale
UNIVERSITY PRESS
New Haven & London

Published with assistance from the Annie Burr Lewis Fund.

Yale University Press books may be purchased in quantity for educational, business, or promotional use. For information, please e-mail sales.press@yale.edu (U.S. office) or sales@yaleup.co.uk (U.K. office).

Set in Bulmer type by Tseng Information Systems, Inc., Durham, North Carolina.
Printed in the United States of America.

Library of Congress Control Number: 2020952843
ISBN 978-0-300-23610-1 (hardcover : alk. paper)

A catalogue record for this book is available from the British Library.

This paper meets the requirements of ANSI/NISO Z39.48-1992 (Permanence of Paper).

10 9 8 7 6 5 4 3 2 1

For experimental gardeners everywhere, especially my mum.
This is for you, Carol.

Contents

Acknowledgments

BOOKS ARE COLLABORATIVE processes that take years to complete and are molded through thoughts, ideas, and comments from the wider academic community. This book is no different. It has also been developed across several different higher-education institutions and while I've been employed in a range of academic as well as one academic-related post, reflecting the twenty-first-century precarity of many postdoctoral careers. So I would like to begin by an admission that this list of acknowledgments is likely to miss many people who have asked questions, offered suggestions, and generally been supportive as I have moved across the country and between institutions, but I am very thankful to you all.

This book owes its very existence to the generosity of the Wellcome Trust, which funded an early pilot study via a small grant in 2011 (095110/Z/10/Z, for the project "Laboratory and Spectacle"), and then a Wellcome Postdoctoral Fellowship in Medical History and Humanities in 2013–2016 (100388/Z/12/Z), which allowed me to develop those early ideas more fully. I was also lucky in that when I was writing the fellowship proposal, I was employed in the Humanities Division at the University of Oxford as a research facilitator and was generously supported by a team of expert colleagues, including Aileen Marshall-Brown, Fiona Groenhout, Alyson Slade, and Sam Sneddon. Once the fellowship began, I received excellent mentoring and support from those in the History Department at King's College London, in particular Anna Maerker and Abigail Woods. This continued when I moved my fellowship to the History and Archaeology Department at the University of Chester and began my first lectureship post. The Faculty of Arts and Humanities at Chester supported final crucial research trips to chase references, and sabbatical time to write the bulk of the chapters. I am particularly thankful to Rebecca Andrew, Tim Grady, Jennifer Hillman, and Caroline Pudney for commenting on chapters and providing intellectual and personal support, as well as the amazing Kara Critchell, who has encouraged and motivated me throughout, as well as casting her critical eye over several chapter drafts. I am now finishing the process in a new post as senior lecturer in history at Newcastle University, where I have already found positive support from new colleagues there, including Sophie Moore, who has cheered me on via our virtual office coffee breaks, and Vicky Long, with whom I have been gigging and drinking since my early days as a PhD student. Other colleagues and friends who have given guidance and patiently listened to me expound on this subject are numerous, but special thanks go to: Victoria Bates, Sarah Bell, Michael Brown, Marianna Dudley, Linden Groves, Alexandra Harris, Claire Jones, Valerie Joynt, Becca Lovell, Alice Marples, Laura Mayer, Elaine Mitchell, Karen Moore, Glen O'Hara, Chris Pearson, Neil Pemberton, Rebecca Preston, Jenifer White, Tim Reinke-Williams, and Rebecca Wynter—many of whom have read sections and all of whom have engaged with, challenged, and thereby helped develop my ideas. I have without doubt been very fortunate in the colleagues and friends I have found myself surrounded with over the last few years, and the book could not have come to fruition without them.

There have also been teams of archivists and librarians who have answered queries, discussed ideas, and suggested lines of inquiry. Specific thanks go to Leonie Paterson at the Royal Botanic Garden Edinburgh Archive, Stephen Harris at the University of Oxford Herbaria in the Department of Plant Sciences, and Eve Watson at the Royal Society of Arts Archive. Appreciation goes as well to the Royal Society Archive, the Linnean Society Archives, Royal Botanic Gardens, Kew Archive, and of course the Wellcome Library and Collection with its amazing and truly dedicated staff, who have been assisting my work since my undergraduate days.

Warm thanks are due to the following institutions for permission to use images within this text: Royal Botanic Garden Edinburgh, Royal College of Surgeons, the Royal Collection, Bridgeman Images, National Gallery of Ireland, Wellcome Collection, Metropolitan Museum of Art, Lewis Walpole Library, British Library, and Yale Center for British Art.

Thanks also go to the peer reviewers who have opened new venues for investigation and generously engaged with the manuscript, and to all those at Yale University Press who have efficiently and diligently worked on making this text the best it could be. Any mistakes or inaccuracies that remain are my own.

I have many personal debts, as well as the academic ones, but Carol and Guy Hickman have been my true rocks of support. My mum, Carol, in particular, has read all of this work at least once, often several times, which goes well beyond the call of familial duty, and my brother, Guy, has managed to keep my feet on the ground as well as act as my personal cheerleader. Without them, it is unlikely that any of this research would have made it out of my head onto the page.

Quick Guide to the Key Medical Practitioners and Their Gardens

These are listed in order of appearance in this book:

JOHN COAKLEY LETTSOM (1744–1815)
Physician and Quaker. He purchased the land that became Grove Hill in Camberwell, three miles to the southeast of London in 1779.

JOHN HOPE (1725–1786)
Physician, botanist, and lecturer at the University of Edinburgh. He created the Leith Walk garden in Edinburgh as a teaching and research space in 1763. This formed an important training ground and model botanic garden for the next generation of botanists and medical practitioners.

JOHN FOTHERGILL (1712–1780)
Physician and Quaker. He purchased Upton House, West Ham, on the outskirts of London in 1762. He was also Lettsom's mentor, and on his death many of his plants were purchased by the younger physician and replanted at Grove Hill.

WILLIAM PITCAIRN (1712–1791)
Physician and president of the Royal College of Physicians. A plant collector who worked with Fothergill, he developed a private botanic garden in Islington, then a village on the outskirts of London, from the 1770s.

WILLIAM CURTIS (1746–1799)
Apothecary. He established the first subscription London Botanic Garden with support from Fothergill and Lettsom in Bermondsey in 1773 (the garden was later moved to Lambeth and then to Brompton). Lettsom also

lent Curtis money to facilitate the publication of his *Flora Londinensis;* the second volume of the *Flora* is dedicated to Lettsom, as well as Curtis's plant-collecting trip to Yorkshire, on the condition that any duplicated specimens would be gifted to him for the Grove Hill garden.

JOHN HUNTER (1728–1793)
Surgeon and anatomy lecturer. He began developing his estate at Earl's Court, a rural village on the outskirts of London, in 1764. This became a domestic laboratory for plant and animal experimentation.

EDWARD JENNER (1749–1823)
Hunter's house pupil, surgeon, and pioneer of the cowpox vaccine for smallpox. He lived at The Chantry, Berkeley, in rural Gloucestershire, where he repurposed his contemplative garden building as a place to vaccinate the poor.

The Doctor's Garden

INTRODUCTION

Illuminating the Doctor's Garden

THE RURAL RETREAT OF Dr. John Coakley Lettsom (1744–1815), at Grove Hill in Camberwell, commanded a panoramic view of the bustling and rapidly expanding city of London on the opposite side of the river Thames. Lettsom described how from his villa, "even the evening scenery presents peculiar beauty; whilst the stars of the firmament form a canopy, the innumerable lights of the metropolis, are extended beneath like a luminous carpet, and pierce the darkness of night with glittering radiance."[1] Here on the urban fringes, with London shimmering in the distance, Grove Hill represented both an escape from the busy work of medical practice in the city and an opportunity for botanic, agricultural, and scientific contemplation for an eminent physician.

Unlike larger and more rural estates, its situation in Camberwell connected Lettsom visually to the city, where his town house and successful medical practice were located, and thereby the commercial activity as a physician, which had allowed him to purchase and develop his "terrestrial Elysium." His professional metropolitan career began in 1770 when Lettsom became licentiate of the Royal College of Physicians and began practicing in Basinghall Street in the City of London. In this year he also married Anne Miers, a union that brought with it a considerable financial settlement.[2] The purchase of the estate at Camberwell in 1779 then represented a statement of personal and professional success and a suitable retreat for a late Georgian medico-gentleman (plate 1). This visual relationship to London also denoted Grove Hill's links to a world beyond the garden gate, and it is this wider context that will be the focus of this book.

Rather than concentrate on an individual garden or a group of gardens affiliated with a single designer, style, or location, this book will place Lettsom's country estate at the center of a medico-botanic network that was connected by the movement of ideas, people, plants, animals, and objects.[3] By repositioning our focus away from a traditional scientific center of calculation—such as the Royal Collection at Kew gardens, also on the fringes of London, and which was similarly being developed into an internationally important scientific resource by Joseph Banks, president of the Royal Society and a keen botanist himself—this book will aim to show how less elite gardens developed and owned by medical practitioners and their networks also played a crucial role in the development and exchange of knowledge during the late Georgian period.[4] Each chapter will begin with Lettsom and his garden before meandering through the metaphorical shrubberies to explore other gardens, collections, and experiences. As well as moving the network off-center, the garden in its various forms as public, institutional, and private will be foregrounded and considered in relation to other spaces of knowledge creation, such as the museum and the library. This is not to say that these gardens were entirely original in their conception as scientific and experimental spaces. In many ways they were following a strong local tradition that included the diarist, writer, and keen horticulturalist John Evelyn's seventeenth-century garden at Sayes Court, in Deptford, London, which aimed to develop and promote experimental knowledge through the inclusion of an "elaboratory" alongside a library, repository, orchard, gardens, and an aviary.[5]

These men can also be seen as following a tradition of collection building, particularly in relation to museum creation. Many of the founders of the early museums in England, for example, were scientific and medical men, such as Hans Sloane, physician and naturalist, whose private collection formed the basis of the British Museum in the 1750s, and Elias Ashmole, chemist and botanist, who founded the Ashmolean Museum in Oxford in 1683.[6] The gardens in this book can be seen within this broader cosmology of the collecting and sharing of knowledge, which gained further popularity in the eighteenth century.

Coalescing around a small network of medical practitioners at the top of their professional field in the Georgian period, most notably Lettsom,

John Fothergill, William Pitcairn, John Hunter, and Edward Jenner, this book will focus in particular on the metropolitan rural villas situated on the fringes of London. Such sites allowed practitioners to maintain successful medical practices with wealthy clientele in the urban center, as well as rural retreats that were accessible when they found time for leisure. However, it will also consider the Leith Walk botanic garden in Edinburgh; William Curtis's botanic garden, which moved through several sites in South London; and Edward Jenner's considerably more rural base in Gloucestershire, as these also formed nodes within Lettsom's network. By considering the role of gardens in terms of their uses and experiences, it will trace their importance in the formation of the late Georgian medical professional—from the construction of botanical knowledge as a student, to an imagined space for retirement at the end of a successful career. In this way the interconnectedness between public, private, and institutional gardens within local, national, and global networks will be made visible.

John Coakley Lettsom and Grove Hill

Grove Hill provides the ideal point from which to pivot and consider the different uses of gardens by a small, connected group of medical practitioners in this period. As an estate, it contained the full range of features related to the expression of polite knowledge that might be found in a rural retreat at this time—a museum, library, botanic garden, designated space for agricultural experiments, classical statuary, and a range of garden buildings, including an astronomical observatory.[7] As will be discussed, these all reflected, albeit on a much smaller scale, the elements and interests of George III and Queen Charlotte at Kew, Richmond, and Windsor. Grove Hill can then be viewed as a particular form of fashionable villa, which combined the scientific with the classical, and the farm with the garden. The reference to the "picturesque" nature of the landscape was made several times within Lettsom's own account of Grove Hill. The use of this term places his garden within the fashionable aesthetic approach of the time. With its picturesque objects of statues, a cottage, and an observatory designed on the model of a classical ruin, it followed the growing taste for the picturesque that emerged in the 1760s.[8] As Michael Symes has described, although most writers have focused on the debates surrounding the visual nature of the picturesque, this was also a style that would have been approached via the other senses, including sound, touch, and smell, as well as having other cultural meanings.[9] This book will also consider these other, often overlooked factors in relation to gardens of this period.

Unusually for a garden owned by a medical practitioner, we also have a wide range of sources available, which allow us to reconstruct Grove Hill in detail in relation to its owner, although these are still more limited in archival range than those associated with other more elite landscapes. However, they include two guidebooks written by Lettsom himself, published in 1794 and 1804, with a plan and engravings, an illustrated poetic description, and accounts by various visitors to the garden. As this suggests, Lettsom was a highly successful medical practitioner who represented the height of medico-gentility of the period. As such, his garden has been described in forensic detail by the historian Penelope Hunting, as has William Curtis's London Botanic Garden and its various locations in South London by Kath Clark. However, other similar gardens, with even less extant primary source material, created by practitioners such as Pitcairn, Fothergill, Hunter, and Jenner, have until now been relegated to a sidenote in broader histories of their medical accomplishments.[10] This work will bring these examples together to consider how, as a group, these men transformed, experienced, and used the landscape, and consider what that tells us about botany, medical practice, and scientific horticultural and agricultural endeavors during the late Georgian period.

In 1815, *The Gentleman's Magazine* obituary for Lettsom was extensive and, as was to be expected for such a successful physician, covered the numerous publications and achievements of the Quaker doctor.[11] His identity as a Quaker was clearly important in terms of his networks and philanthropic activities, although perhaps not entirely limited by this identification. It can be argued that in terms of gardens and their use, his role and network as a medical professional were more significant than his religious persuasion, although this was still important. The obituary, for instance, also described how he "in many instances, fostered genius, cherished science, and expanded the circle of the arts, in periods of individual and national distress unprecedented in the annals of this country; and his *purse,* equally with his *pen,* were devoted to their cause. Medicine and botany were particularly indebted to his zealous researches."[12] Among his many accomplishments they noted his release of enslaved people on the plantation he inherited as a young man, his involvement in the Royal Humane Society, his instigation and founding of the Medical Society of London, and his establishment of the General Dispensary in Aldersgate and the General Sea-Bathing Infirmary at Margate. Related to this long list of his achievements, and in particular his interest in science, botany, and medicine, was the creation of the villa and gardens at Grove Hill. This was considered important enough to be noted in the obituary, where reference was given to the improvements he had made to the site as well as the creation of a botanic garden, natural history museum, and library, all of which were described alongside his other professional achievements.

Once he had built up his successful London-based medical practice, Lettsom bought a tract of land in 1779 in Camberwell, then a rural village on the outskirts of the metropolis. Here he built a villa, Grove Hill, and developed a landscape for pleasure as well as botanical and agricultural experimentation. As a leading physician of the period, he could afford a town house near his practice in the city as well as this small rural estate.[13] Although landscape historians have often overlooked small villas such as this, located on the urban fringe, Sarah Spooner has highlighted how they were a common and important feature of the eighteenth-century urban periphery.[14]

Like medical men, writers and artists also had similar desires to be close to their London town houses. In 1753 this need led to the playwright and actor David Garrick purchasing the villa Hampton House along the Thames for a range of activities, including haymaking and garden parties or fetes, which were illuminated by numerous lamps.[15] This mirrored many of the activities that took place at Grove Hill and underlines the importance of a peripheral location for easy movement to the city for business, and from the urban center to the rural retreat for entertaining. As Jon Stobart has discussed, a growing number of elite families in late eighteenth-century

London were increasingly attracted to the suburban villa. This was because these were places that were convenient for, yet also somewhat removed from, the city.[16] Stobart questions the role of these villas and asks whether we should consider them as attractive to the eighteenth-century commuter because of their convenience to the city, with its commercial and social life, or whether we should perhaps conversely perceive them as retreats or places of "polite retirement."[17] In the case of Lettsom and other medical practitioners in this book, I would argue that the semirural estate allowed for both of these functions, so the villa was at once convenient for conducting a busy urban medical practice as well as a place for quiet relaxation and contemplation. Therefore, a focus on elite medical practitioners allows us to extend our knowledge of these smaller but no less significant places. It also situates them in communication with other types of urban gardens, such as subscription botanic gardens and university gardens, which also reflected the growing urban sensibilities of the period. This neither urban nor totally rural location also reflected shifting identities for this group of well-connected medical practitioners, whereby they attempted to balance their gentlemanly position with a growing professional status. Like the situation of their gardens, they were positioned with one foot in an older rural tradition and the other in the growing class of urban professionals.

The time period of this book, which is loosely framed by the reign of George III (1760–1820), covers a remarkable period in British botanic collecting. In 1779, when Lettsom started building his villa and improving his estate, Britain was at the height of plant collecting and interest in botanical subjects. By this time Kew gardens already had around fifty-five thousand species of plants, a number that doubled again by 1814.[18] The combination of the introduction of the Linnaean classification of plants and the new introductions brought back via voyages, such as that undertaken by Captain Cook and Joseph Banks to the South Seas in 1769, made botany a fashionable occupation. As Richard Drayton explores, the personal interest of George III in useful animals and plants, and expansion at Kew, were at the forefront of this development, with the king seeking to fashion himself as the "empire's first gentleman, the paradigm of an 'improver.'"[19] As these elite medical practitioners sought to achieve social approbation and entry into the circles of the gentry associated with Court, public demonstrations of culture, taste, and polite knowledge—including botany and agriculture—were part of the fashioning of their professional identity.[20]

Rethinking the Garden

There is a vast literature on the late Georgian garden. However, until recently this has been predominately organized in relation to specific garden studies, particular designers, wealthy landowners, time period, or regional identity.[21] As Spooner makes clear, "By focusing on the same people and places, garden historians have tended to obscure the variety and complexity of designed landscapes in eighteenth-century England and to simplify the complicated chronology of change."[22] The focus of this book, however, on the gardens associated with the emerging professional class of medical practitioners—which includes physicians, surgeons, and apothecaries—is part of a growing trend in garden history, which has recently included the exploration of alternative landscapes, beyond the elite eighteenth-century examples, and particularly the English landscape gardens associated with named designers.[23] As Malcolm Dick and Elaine Mitchell argued in 2018, in recent years there has been a move away from a focus on aesthetics and design, toward histories of gardens and gardening that are based more completely within their wider social, economic, political, and cultural contexts.[24]

The attention to the use and experience of gardens rather than a focus on their aesthetic design also allows for new interpretations of landscapes. Again, as Dick and Mitchell argue, "Gardens can be understood as part of the shaping of urban and rural landscapes and are influenced by scientific, technological, industrial, medical and intellectual developments."[25] This study offers a new way of thinking about gardens created by a professional class, which were used for a range of scientific, medical, and sociable activities. This approach is best exemplified by recent work concerned with nurserymen, public parks, asylums, hospitals, schools, and public houses.[26] One common obstacle faced by the historian when trying to consider how these places were used and who might be involved in the activity of gardening is the lack of extant archival material.

As this suggests, there are particular challenges associated with the study of landscapes of the emerging professional classes. Tom Williamson has described the new garden history as a multidisciplinary approach in which investigation of the physical fabric is ideally combined with written documents, maps, and illustrations.[27] This is a model methodology for landscape history, but difficult to achieve when studying gardens that rapidly change hands and/or are physically built over. In the case of many of the gardens explored in this book, the rural retreat or botanic sensorium is now subsumed under the streets of expanding urban centers. Similarly, the people involved in the creation and maintenance of such gardens, from professors of botany to day laborers, are also much harder to locate within the archives. Easterby-Smith's work on eighteenth-century nurserymen has offered one model for how to successfully reconstruct the networks, scientific activity, and influence of less elite groups without extensive archival resources.[28] Likewise, by shifting the focus from the visual appearance of the gardens and who designed them to a study of use and experience, it is hoped that a different cultural history of the garden can emerge.

Building on Kate Felus's recent publication, *The Secret Life of the Georgian Garden,* this book will focus on the "use and happenings in the garden."[29] Part of the inspiration for this approach came from my participation in a workshop held at King's College London by Alice Marples and Victoria Pickering in 2015 on the subject of "Collections in Use."[30] This orientation toward use rather than design or creation also puts the garden as a collection of plants into a clearer relationship with other collections, such as museums and libraries, and builds on the work of scholars such as Paula Findlen and Arthur MacGregor, who have outlined the establishment of early modern museums and as part of wider natural history collecting and aesthetic garden practices.[31]

It also allows for a greater concern with those who were maintaining the gardens, the all-important gardeners, as they were fundamentally part of the movement of people between spaces, and for other nonbotanic introductions, such as animals, which may otherwise be overlooked unless they were located within a designated area, such as a menagerie or a model farm. The consideration then of the *use* of the garden, rather than its designed features, also allows for a collapsing of categories of space, such as the farm, ornamental garden, kitchen garden, and menagerie. It allows us to notice, as just one example, the tortoises kept by Lettsom at Grove Hill in the kitchen garden and fed by the gardeners—an example that does not fall easily into traditional garden history categories.

Crossing between the Histories of Medicine, Science, and Environmental History

The new garden history as noted above is now a multidisciplinary endeavor that uses a range of methodologies and approaches to try to bring understanding to the broader context in which designed landscapes are constructed and used. As with the trend toward a concern with less elite gardens, there has also been a move to consider the designed green space within its broader environmental context. In 2018 a special issue of the journal *Environment and History* dedicated to parks and gardens included an editorial penned by the editors Karen Jones and James Beattie. This noted the growing importance of bringing garden and environmental histories into conversation with each other.[32] By taking a broader view of a range of gardens and considering their roles within wider concerns regarding botany and agriculture, science and medicine, this book hopes to add a new perspective to this dialogue between environmental and garden history. This approach has been inspired by key works pioneering this methodology, most notably Mark Laird's *A Natural History of English Gardening,* which offers an environmental view of gardening and its connections with other forms of natural knowledge creation.[33]

The garden as a scientific space is another emerging category of analysis. Such works as the collected volume of essays *Gardens, Knowledge and the Sciences in the Early Modern Period,* edited by Fischer, Remmert, and Wolschke-Bulmahn, speak to this recent development in the history of science.[34] However, the main focus of their work is the contributions of sciences such as mathematics and botany to early modern garden designs and culture, rather than the relationship to medicine and medical practice. As Paula Findlen suggests, in her work on early modern Italy, the material work of garden creation was reliant on numerous naturalists, mainly medical practitioners such as physicians and apothecaries, as well as custodians and owners of botanical gardens.[35] This is equally true of late Georgian gardens and gardeners in Britain, where written texts describing detailed scientific activity are often absent, but it is clear from the material evidence that the physical gardens they created and maintained formed the material base of a wider scientific network.

Horticulture as a scientific practice within the confines of such gardens has been highlighted by Easterby-Smith, who stresses that the cultivation of a garden for such owners was not just the end goal in itself, but that it also provided a place within which further botanical knowledge could be acquired.[36] This focus on a concern with the places where science happened builds in particular on the work of historical geographers, which has highlighted the relationship between place and scientific practice. David Livingstone's *Putting Science in Its Place* notably included botanic gardens as key spaces of scientific endeavor, and Paul Elliott has written several works concerned with parks and gardens and their associations with both the Enlightenment and nineteenth-century science.[37]

Studies of botanic gardens, for perhaps obvious reasons, have been at the forefront of this approach. Emma Spary's *Utopia's Garden* deftly weaves together the history of the Jardin du Roi in Paris, the equivalent, to some extent, of Kew as it was also a royal botanic garden, with its network of scientists and gardeners and the wider social and cultural context of natural history in France. This provides a clear model for the consideration of gardens as scientific and cultural entities that are created, maintained, and remade over time.[38] Other work that places botanic gardens in conversation with each other and that maps their change in use and meaning over time have also been influential in helping to illuminate the range of gardens encountered and developed by medical practitioners in this period. In particular, Therese O'Malley's interrogation of the relationships between design, art, and science in botanic gardens in the long eighteenth century, and Nuala Johnson's comparative work on the botanic gardens of Dublin, Cambridge, and Belfast have been influential.[39]

The approach of this book builds on this scholarship and also corresponds to the growing interest in the development and practice of science beyond that of the nineteenth-century construct of the laboratory.[40] However, unlike recent work in this area by Robert Kohler, Simon Naylor, and others, this work will focus on designed spaces rather than the field and therefore also seek to redefine the garden as a "liminal" space that exists between the wilder "field" and the more managed and ideally placeless "laboratory." In many ways this reflects the move toward considering the domestic as an important environment for the development of scientific ideas and practices in the early modern period, as denoted in particular by the volume *Domesticity in the Making of Modern Science,* edited by Opitz, Bergwik, and Van Tiggelen.[41]

This book instead takes a particular group of professional gentleman as its focus—British medical practitioners. These men were at the heart of botanic endeavor in the eighteenth century, and the gardens, which they created and re-created, acted as important nodes of botanic and scientific inquiry. These spaces were also important sites for the circulation of knowledge, objects, and people and formed part of much larger national and global networks. As Londa Schiebinger states, "Science followed trade routes at the same time that naturalists—in the eighteenth century, mostly physicians—worked to improve commerce."[42] Medical practitioners then were key figures moving between places and helping transport both knowledge and natural history specimens, as well as developing places for botanical experiment and enterprise.

In this way they also created spaces that formed part of global imperial networks, and their collection of plant and other material from around the world meant that they were entangled in the trade of humans and other cargo, as well as the exploitation of indigenous knowledge and resources. Although this is implicit rather than explicit in the accounts of the gardens created, used, and experienced by medical practitioners in Britain, it is worth noting that gardens are not created outside the economic realities of the time but are rather embedded within them. As Beth Tobin argues, the "aesthetic rendering of cultivated landscapes and the abundance of the market place contains and obscures the unsettling economic relations of exploitation that undergird such images."[43] Correspondingly, as Tobin,

Drayton, and Miles Ogborn in particular have shown, by moving beyond narratives of aesthetic beauty and design in the garden, we can unpick the broader relationships between gardens on a global scale and thereby connect them to the world beyond the perimeter wall.[44] This reconnection of gardens to their broader context also helps to reveal elements that are often hidden in relation to garden creation and maintenance, including the labor of women and gardeners, as well as the interlinking of the arrival of exotic plant material via slave-trading routes.

This book will focus on a small group of regular medical practitioners, as opposed to irregular practitioners such as quacks and drug peddlers who were also involved in the medical marketplace. In this period, regular practitioners fell into three main categories—physicians, surgeons, and apothecaries, although there was much fluidity between these titles.[45] For example, some practitioners, like Lettsom, began as apprentices to apothecaries before qualifying as physicians. However, it is fair to say that physicians were at the top of this triumvirate structure in terms of the levels of pay they could command and the polite social circles in which they could move. Therefore, the gardens within this book were created predominately by this group—hence the title *The Doctor's Garden.*

Orientation Guide

These gardens are multifaceted and there are many different methods and approaches by which they can be examined. For the purposes of this book, I have applied six lenses, all of which aim to highlight the ways in which the landscapes were designed and used, as well as the ways in which they were connected to other similar spaces. Chapter 1 focuses on the roots of botanic knowledge within the medical curriculum and considers the use of gardens for education, sensory approaches to knowledge creation and dissemination, as well as the importance of trained gardeners as botanical assistants. The second chapter considers the role of domestic gardens developed by doctors in Britain as nodes within local, national, and global networks and the interrelationships between science, sociability, pleasure, and the senses. Chapters 3 and 4 consider the reception of the gardens by visitors and how they may have been understood and read using a range of guidebooks, labels, and other texts. The blurring of lines between the farm and the garden are identified in chapter 5, and chapter 6 begins with a garden party and moves on to an exploration of the sociable nature of such places. To round things off, the book concludes with an epilogue that considers how this historical approach to thinking about the use and experience of gardens could be applied in heritage interpretation, thereby offering fresh insights into the bridging of theory and practice.

ONE

Educating the Senses

The Botanic Garden as a Teaching and Research Center

JOHN COAKLEY LETTSOM as a teenager would have been found trawling the fields, woods, and hedgerows of Yorkshire closely observing and collecting the plant material native to the region. We know this, like so much about Lettsom's life, from Thomas Pettigrew, the early-nineteenth-century surgeon and antiquarian. He published an extensive three-volume biography of Lettsom, including reproductions of his correspondence, in 1817, just two years after Lettsom's death.[1] This provides a detailed background to the elder physician's life and times and includes an account of the emergence of a life-long interest in botany from Lettsom's time as an apothecary's apprentice in Settle, Yorkshire, before he began his training as a physician. In the second year of his apprenticeship, when he was around nineteen years of age, Lettsom had already developed a love of botany over other forms of study.[2] Pettigrew describes how, in order to develop his skills in this area, "he borrowed Gerard's Herbal and in his excursions in the vicinity of Settle, he collected many good specimens of rare plants, with which he composed an Hortus Siccus," or dried collection of plant specimens, otherwise known as a herbarium.[3] Here we can already see a youthful interest in plants, being shaped and developed through his training as a medical practitioner.

This collecting of plants on botanizing or herborizing excursions into the countryside and then drying the specimens reflected the activities of the metropolitan apprentices of the Society of Apothecaries, who were taken on similar but larger group training expeditions in and around London (fig. 1.1). These were undoubtedly livelier affairs than those conducted by a solitary Quaker apprentice such as Lettsom. While he was exploring the flora

Fig. 1.1. William Curtis and friends on a botanizing expedition, as depicted on the frontispiece to Curtis's *Flora Londinensis,* which cataloged and described the plants found in the London area. Stipple engraving by W. Evans, 1802. Wellcome Collection, CC BY 4.0.

of Yorkshire on his solo adventures, the Court of the Society of Apothecaries was instead dealing with complaints from members regarding "riotous" group herborizing expeditions due to the activities of the apprentices.[4] During the 1760s many of the society's members were discouraged from the tradition of sending their apprentices on botanical expeditions organized by the society, due to the "irregular and indecent behavior" of those who treated it as a holiday rather than a learning experience.[5] Because of such behavior the rules were soon changed to reduce the amount of socializing involving alcoholic refreshments in local hostelries, which was enjoyed by apprentices.

The herborizings of the apprentices in London seem almost tame in comparison to descriptions of Linnaeus's explorations beyond the classroom. In Robert Hunter Semple's 1878 *Memoirs of the Botanic Garden at Chelsea,* he relates the following in which the herborizing activity was linked to the botanic garden by the practice of walking between the two:

> In Linnaeus's Diary we are informed that that distinguished Professor, during his summer lectures, took out with him about 200 pupils, who collected plants and insects, made observations, shot birds, kept minutes, and having botanized from seven o'clock in the morning until nine in the evening every Wednesday and Saturday, returned with flowers in their hats, and accompanied their leader with drums and trumpets through the city to the garden.[6]

This is altogether more performative than the young Lettsom collecting plants from the wild on his solitary expeditions, and points toward another function of the group activities—that of a more polite social experience wherein professional relationships could be created and affirmed. At Chelsea, one of the summer herborizing expeditions, known as the General Herborizing, was chiefly for members of the society and any invited guests. Here practicing apothecaries, rather than apprentices, could hone their skills and socialize beyond the city walls, apparently finishing the day in a hostelry, where haunches of venison were evident as examples of conspicuous consumption.[7]

Alongside this experience of the field as a space for developing botanic knowledge, the garden base, with its array of rare and interesting specimens, was also a training site. The delivery of botanic teaching in a garden such as the Chelsea Physic Garden for apothecary apprentices in London, or university botanic gardens for medical students, as well as essential botanizing in the wild, relied on the same sensory contact with plant material. In 1777, John Hope, who was eminently respected as the Regius Keeper of the Royal Botanic Garden Edinburgh, king's botanist, physician, and lecturer, introduced students to his annual course on botany in his state-of-the-art botanic garden on Leith Walk, Edinburgh (plate 2). As the study of plants formed the focus of the course, botany was traditionally taught within the botanic garden itself, and Leith Walk was no different. What was certainly

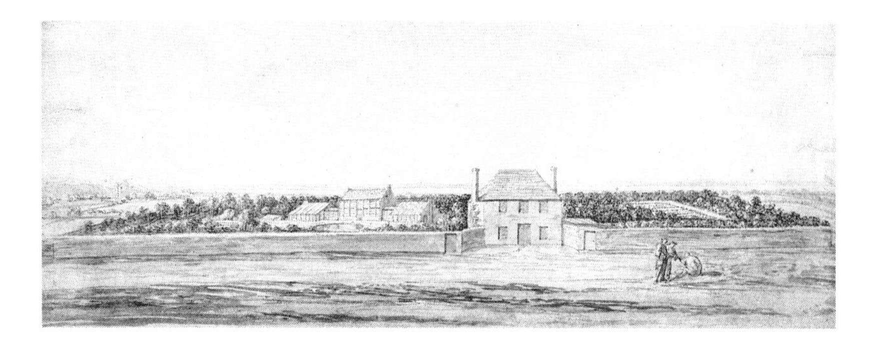

Fig 1.2. Perspective view of the Leith Walk garden, by Jacob More in 1771, with the botanic cottage located contiguous with the perimeter wall. Within the garden can be seen greenhouses on the left, and in the foreground on the road outside, physicians are depicted helping a sick man, to highlight the medicinal nature of the space. Reproduced with permission of the Trustees of the Royal Botanic Garden Edinburgh.

unique at the time was the layout of the five-acre botanic garden on the outskirts of Edinburgh (fig. 1.2 and plate 3). As Henry Noltie has stated, this not only included the botanic cottage, which encompassed lodging for the gardener and a teaching room, but also a 140-foot-long set of conservatories that contained a 70-foot-long greenhouse at the center and large hothouses at either end.[8] These were all set within a landscape of winding paths running between beds of exotic as well as native collections of plants. To anyone traveling along Leith Walk in the late eighteenth century and catching a glimpse of exotic trees reaching above the walls, it must have seemed an extraordinary and otherworldly place.

University botanic courses in this period were open to any man who could afford to pay the fee (all funds were paid directly to professors for their lectures), but the main audience was predominately medical students. At Edinburgh, botany was a compulsory subject from 1777, along with anatomy and surgery, chemistry, materia medica and pharmacy, medical theory and practice, and the clinical lectures of the Royal Infirmary.[9] This is not to say that all medical students attended botanical lectures. At Edinburgh, for example, Lisa Rosner suggests that only around 25 percent of each cohort was present at the end of the eighteenth century.[10] This was partly because the course ran in the summer months, when students wanted to be back home, but it was also not the most sought-after skill. According to the *Guide for Gentlemen Studying Medicine at the University of Edinburgh,* penned under the pseudonym of J. Johnson, the main reason for studying botany was that it would prevent the practitioner's abilities from "being called into question by his ignorance of the principles of a science which is vulgarly believed to be necessary."[11] As with all student feedback, this is only one opinion, and Noltie's research suggests that John Hope's lectures on botany at least were generally well attended, if not by medical students. The range of students in his classes went well beyond those with medical interests, suggesting a shift away from medicinal botany to botanical study in its own right, as well as a growing interest in botany for agricultural and other economic purposes.

This garden provided the necessary specimens for Hope to conduct a sensory approach to the teaching of botany. In the introductory lecture of his course on botany, he outlined his belief that "we derive more knowledge from the senses viz. the taste and smell, than from all books together. I can say thus far for myself, that I got more knowledge from these in the Materia Medica, than in books."[12] The role of all the senses in both learning and understanding plants seems to have been of crucial importance to Hope. The "Materia Medica" he was referring to was the term given to the substances (vegetable, animal, and mineral) prescribed by medical practitioners to treat disease. He evidently considered that through an investigation using all his senses he learned more about these botanic specimens than from written descriptions printed in books. The senses were then given a higher authority than knowledge transmitted through written texts by Hope, thus highlighting the importance given to the role of physical plant material in the understanding of plants in terms of their physiology, anatomy, and therapeutic effectiveness.

This chapter, then, will explore the role of the senses in botanical knowledge dissemination and how this related to the creation and use of botanic collections as well as a wide range of other media in teaching, such as illustrations, dried specimens, and other objects. Focusing mainly on Hope and the Leith Walk garden in Edinburgh, it will also consider the problems encountered when trying to teach botany without a well-resourced garden, as demonstrated by the example of the University of Glasgow. Although botanic gardens in this period are often considered by historians as loosely related but separate spaces to general medical teaching, this chapter will also look at the pedagogical methods shared by anatomy and botany and use the example of Andrew Fyfe, illustrator, botanist, and anatomist, to explore the common nature of these disciplines when the lens is focused on the science of the classroom.

Hope's Botanical Sensorium

Human interaction with plants is likely to have always been multisensory. We have eaten them for pleasure, enjoyed their perfume, appreciated their different textures, listened to the sound of the wind through their leaves, and of course been delighted by their immense visual variety in shape and color. During the long Enlightenment, these sensorial engagements in the garden were enhanced by the introduction of new exotic species from overseas. From the seventeenth century, gardens aimed to achieve maximum multisensorial pleasures, and this was achieved through the importation and growing of exotic floral plants, such as highly perfumed jasmine.[13] Similarly, the great palatial estate at Versailles in Paris was constructed to arrest all the senses through its multitudinous brilliant gold fountains, grottos, and statues, and the sheer size of both the chateau and the gardens, which were designed to dwarf the visitor.[14] The gardens were, of course, also the backdrop to other sensory spectacles, including moving water, dramatic illuminations and fireworks, music, dancing, plays, and poetry.[15] This demonstrates the prominence that "sensing and feeling" were given during the Enlightenment in relation to gardens, and so Hope's call to use all the senses in order to know about plants can be seen within this cosmology of sensory knowledge creation and experience.[16]

Hope can also be situated within a long tradition of botanists who developed their knowledge and understanding using the whole range of senses. Elements such as taste and odor had been mentioned in herbals and natural histories reaching back to at least the ancient Greek physician and botanist Dioscorides.[17] Attempts to define these ephemeral senses in relation to plants became part of new taxonomic approaches during the Enlightenment. As Mark Jenner has described in relation to the late-seventeenth-century Lichfield physician John Floyer, taste might be utilized as a method by which the therapeutic benefits of plants could potentially be ascertained and organized. Even though in the case of Floyer his approach to tasting a wide range of substances (including mineral, animal, and vegetable) might result in the cataloging of tastes that encouraged complexity, contradiction, and confusion.[18] Despite this, in his published work, *Pharmako-basanos,* Floyer argued that there was "no Vertue yet known in Plants, but what depends on Taste and Smell, and may be known by them."[19]

Jenner describes how this worked in practice, using the example of Floyer's visit to the Chelsea Physic Garden in London in the 1680s, where Floyer notes that he was "pleas'd with many Curiosities" and remarked on the "Ingenuity" of the ordering system used by the society.[20] While there, he also admired the great number of specimens in the garden and tasted many of them. Jenner argues that "this was not mere recreational grazing: Sir John Floyer was chewing experimentally. He recorded his perceptions of taste and smells having conferred with the keeper, Mr Watts, whose 'Taste and Smell did very much agree with Mine,' and with his companions, the London Collegiate physicians, Edward Baynard and Edward Betts Junior."[21] This was an approach wherein agreed distinctions between different sensory perceptions were attempted through discussion with other trained practitioners, as well as one's own experimental chewing.

Hope seems to have followed a similar methodology concerning the importance of smell and taste in his lectures. According to one set of student notes, he outlined the importance of senses such as smell in animals and used the example of dogs knowing their master through scent alone. He also argued that in order to understand plants, botanists needed to keep their senses as sharp as possible, stating:

> The person who is to make experiments on taste must be in a state of health, & the organ no way affected by any other thing & it must also as I said be in a state of health, & the tongue must

> be moist not dry. He who means to make experiments on plants, must live as temperate as possible. We see in common life that a person who tries wine by the taste tries it in the morning, because at this time his organ of Taste is most acute he must be also accustomed to live on simple food, the more accurate the taste will be; we must also exercise this organ.[22]

Here the sense of taste is described as a way of knowing that should be nurtured and trained. The senses should also be unpolluted by the more luxurious aspects of modern life, in this case rich food and alcohol, which could dull their investigatory powers.

However, a reliance on these more nebulous senses could also be problematic. In a system such as Linnaeus's taxonomy, which was based on the visual recognition of the reproductive parts of plants, other sensory ways of knowing began to take second place to that of sight. This is not to say that the scents of plants along with their colors and flavors were omitted from descriptions. Linnaeus himself claimed that "all the world's plants emit one of five odors: ambrosial (such as amber and musk), fragrant (like jasmine), spicy (like sassafras, camphor, and citrus fruit), noisome vegetable scents (like cannabis or opium) or nauseous vegetative scents (like tobacco)."[23] However, as Holly Dugan reminds us, Linnaeus explains to "his readers that 'scent never clearly distinguished a species,' since the sense of smell is 'the most obscure of all the senses' and because the scent of all things very easily varies."[24] Scent, then, was notoriously difficult to pin down into an ordered system for botanical knowledge.

Similarly, William Cullen (fig. 1.3), a Glasgow-based physician and professor, shared Linnaeus's concern about the limitations of various sensory ways of knowing. In 1761 he argued that in order to discover the medicinal virtues of plants, "*Colour,* of all methods of knowing the virtues of the subjects *a priori,* is the most uncertain; *Smell* extends a little farther, but *Taste* is the most extensive of all the three."[25] He went on to describe these areas in more detail. He wrote, "With regards to Odours, I find this very difficult, as they are of such infinite variety, and of so little resemblance, as makes it very difficult to reduce them to any general heads, so that thence

Fig. 1.3. William Cullen, physician and lecturer, who conducted agricultural experiments on a farm near Glasgow. Stipple engraving by F. Holl after D. Martin. Wellcome Collection, CC BY 4.0.

we might derive particular virtues from the different kinds of them. Linnaeus has attempted a distinction of this sort."[26]

In relation to taste, he claimed that this sense "labours under the same difficulties as Odour. The perceptions from the same impression vary in Smell remarkably, in Taste considerably so. There is not only the same difference of what is grateful to one being not so to another, but also a difference with regard to impression, what is acrid to me being almost insipid to another."[27] Similarly, in his lectures, Hope outlined to his students the limitations of taste in knowing the effect of plants on the body:

> If they have a sensible effect on that organ of taste they have the same on the system; whether it be permanent or fugacious. This shows we have a large field for substitution but we are not to judge of the effects of plants entirely by the taste, for some that have little or no effect on the sapid organ, have a great deal when taken into the body, yet there are limitations to this Rule.

We can perhaps identify here the beginning of a shift from Floyer's attempts to calculate how senses such as taste might map onto the therapeutic use of materials such as plants. Cullen had certainly moved further away from the idea that the therapeutic impact could be determined by smell by this point. He concluded that "upon the whole, very little of the medicinal powers are to be determined from the odour."[28]

However, even with these limitations the full range of senses could still play a part in ways of knowing, and a multisensory approach to understanding was shared by the broader medical community. The American physician Benjamin Rush told his students that "in a sick room, we should endeavor to be all touch, all taste, all smell, all eye and all ear, in order that we may be all mind; for our minds, as I shall say presently, are the products of impressions upon our senses."[29] This reflected wider beliefs of how the senses created impressions on the mind. As Rousseau wrote in *Emile* in 1762, "To exercise the senses is not only to make use of them, it is to learn to judge well with them. It is to learn, so to speak, to sense; for we know how to touch, see and hear only as we have learned."[30] Similarly, Cullen argued that in order to understand the materia medica, "*the knowledge of the subject* is of two kinds, natural and artificial; the *first* procurable alone by the too much neglected study of Natural History; the *last,* by the frequent inspection, or handling of the subject."[31] This then places the botanic garden at the center of this approach, where all the senses were utilized in the process of learning, and frequent inspection and handling of materials were essential in order to understand the natural world.

Hope's botanical course was one of the most advanced in Britain at the time, both in the breadth of its coverage but also in its use of experimental research, visual aids in the form of teaching posters and other drawings, and arguably the use of the extensive botanic collection within which the course took place.[32] Hope's course of around sixty lectures was held in the summer months and was formed of three parts: "On Vegetation," "On Classification," and "Demonstration." "On Vegetation" explored the anatomy and physiology of plants, which highlighted the scientific approach to botanical knowledge beyond simply the medicinal use of plants. Within this section Hope used his own experimental work on plants as part of the teaching methodology. "On Classification" was based on the taxonomy of plants and based on Linnaeus's work in this field. The final element was a demonstration in which Hope showed students living examples of plants within the garden. These three areas interlinked and blended into each other, and it is clear that the Leith Walk garden was a teaching and research center that was essential to Hope's sensory approach to understanding and teaching botanical knowledge.[33]

University Botanic Gardens as Pedagogic Spaces

Like the use of the senses as ways of knowing in botany, scholarly botanic gardens already had a long history by the time Hope was teaching in Edinburgh. They had started appearing in the sixteenth century in Europe, with early examples at Pisa, Padua, Leiden, and Montpellier, followed closely behind in the seventeenth century with the scientific botanic gardens of Oxford, Uppsala, and the Jardin du Roi in Paris. As today, the reputation of individual universities was heightened by the employment of high-ranking academics as well as the creation of novel teaching facilities, including a botanic garden and an anatomical lecture theater.[34] In the case of the early botanic gardens, they could be arranged both "symbolically and practically," as John Dixon Hunt has identified.[35] At Leiden, for example, the long thin beds were laid out for easy access to the plants and their labels for both teaching staff and students, whereas the circular garden at Padua was divided into four sections representing the four corners of the world, although within that a similar organization of beds designed for teaching was also present.[36] A functional layout that prioritized study suggests that much of the teaching took place within the garden itself, as shown in a 1601

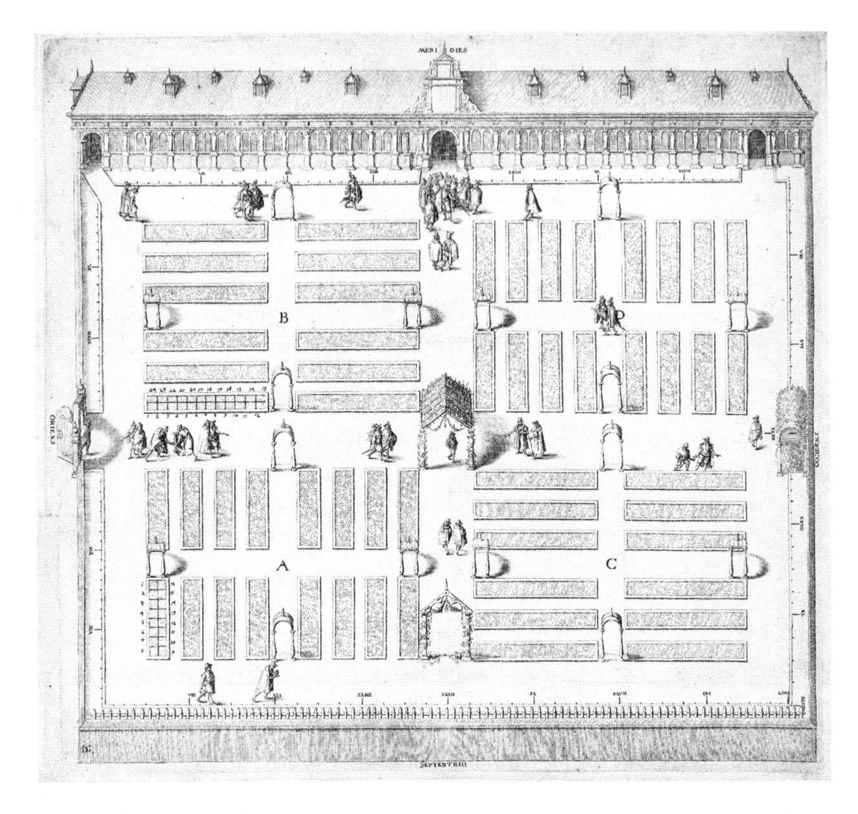

Fig. 1.4. Depiction of the Leiden botanic garden in 1601, with a lecturer teaching a group of students in the top center of the image. Gezicht op de Kruidentuin (Hortus botanicus) van de Leidse Universiteit, Jacob de Gheyn II. Rijksmuseum.

engraving depicting a master and his students, as well as others, in the Leiden botanic garden by Jacob de Gheyn II (fig. 1.4).

When we think of botanic gardens today, we tend to see them as collections of native and exotic plants that have been laid out either taxonomically or to be aesthetically pleasing (or sometimes both). However, medical students have historically formed the main group that have used the botanic or physic gardens associated with universities as part of their training.[37] This is because the aim of the early botanic gardens was specifically to educate doctors in the knowledge of plants related to medical practice.[38] In Leiden, as at Padua, Montpellier, and Oxford, students would have mainly learned how to identify plants and their uses as therapeutic interventions. Known as "simples," plant remedies were an important part of the practitioners' tool kits. Although this is not to say that botanic gardens only functioned on one level. Such gardens were never just about the collection of botanical material; they were also places that housed a range of other natural history specimens.[39] Partly because of their varied collections, they also acted as places for knowledge exchange, as scholars met to handle and discuss the latest discoveries that had been sent back from foreign expeditions, whether plant, animal, or mineral, in a similar way as other sociable scientific spaces such as museums and pharmacies.[40]

During this period botanic gardens developed as scholarly places where people could engage with each other and a variety of material substances. As Findlen argues, "Botanical gardens were not only constructed out of the materials naturalists brought back from their voyages but they became a replacement for travel itself—a laboratory of nature that allowed the observer to absorb the collective medical and botanical knowledge of the age."[41] The growing presence of natural history museums, which mushroomed alongside university botanical gardens, also highlights the growing focus on both demonstration and observation within the Enlightenment medical curriculum.[42] As Hope outlined in his lectures, demonstration and observation were central elements of his pedagogical approach to teaching botany.

Although gardens, specimens, and botanical objects were still used for teaching, by the end of the eighteenth century there was a shift away from the use of botany in the training of physicians in Britain as a predominately functional part of medical training. For example, the *Guide for Gentlemen Studying Medicine* stated that "it has been alledged with some plausibility that the study of botany, in the present improved state of medicine, is not necessary to practitioners of the healing art, as all the medicines which the vegetable kingdom furnishes are found in the shops, and described in every treatise on materia medica."[43] This statement is very telling: it denotes a move from the use of botanical knowledge as an integral part of a physician's medical practice and highlights the increased role of commercial activity in the creation of drugs and the enhanced role of the apothecary in selling them. It also suggests that there was a perception that books on the subject could replace the hands-on medicinal botany training traditionally taught in university gardens.

The *Guide,* however, also indicates that there was a new role for botany in the late-eighteenth-century world inhabited by the enlightened medical practitioner. The author argued that "when it is considered that botany, like other branches of natural history, has now become part of the education of every gentleman, no medical practitioner will choose to hazard his abilities being called into question by his ignorance."[44] Here the writer suggests that botany should be studied so that a physician cannot be caught out or found ignorant of the subject by a layperson or perhaps by a competing form of practitioner, such as an apothecary. The suggestion that it was an essential "part of the education of every gentleman" underscores the growing status of botany as an important form of polite knowledge among the educated and elite classes.[45]

This creation of the identity of the eighteenth-century physician as a knowledgeable gentleman has been traced by Michael Brown. Using his mode of analysis, the botanic garden can be seen as an important space that facilitated the acquisition of botanical and agricultural forms of polite knowledge as well as medically relevant information.[46] The polite nature of botany by this time was further emphasized by the finding that it was not only medical students who attended the Edinburgh botany course. A list compiled in 1763 and published by Noltie indicates that the following paid a fee to attend: "a Knight, 2 ministers, a captain, druggists, advocates, Americans and some noted simply as 'Infirmary.'"[47] Similarly, the course extended beyond British shores. Between 1761 and 1786, seventy-seven students came from North America, thirty-five from the West Indies, and twenty-one from other European countries, such as Switzerland, Germany, and Spain.[48] This broad interest in botany both within and without the medical curriculum and beyond the borders of Britain establishes the new status that the science acquired in this period and its fashionable nature for all forms of gentlemen, including the new professional physician.

Scotland led the way within Britain in developing new Enlightenment modes of teaching, which included botany as a compulsory part of the medical curriculum in Edinburgh from 1777.[49] When botany as a subject became popular on the Continent and medical students who could afford it traveled to gain the most expert training they could find, up-to-date teaching resources, such as well-stocked botanic gardens, became crucial in attracting students. In 1807 the Glasgow faculty minutes noted that in "every well-endowed University . . . the teaching of Astronomy implies an observatory furnished with Instruments, and Lectures on Botany presupposes a Garden of Plants."[50] They were discussing the importance of the eminent surgeon William Hunter's collection of art and objects, including natural history material, which had just been donated to the university for use in teaching (these now form the basis of the Glasgow-based Hunterian Museum), but it is clear that botanic resources, in the form of a garden, were regarded as an essential resource for any decent university, in line with an observatory for astronomy and a well-stocked museum.

As early as 1758, the somewhat infamous actor, botanist, and physician John Hill had responded to the growth in numbers of students going overseas for university botany courses by arguing that London needed to establish a botanic garden in order to keep medical students at home rather than traveling to the Continent for training. He felt that there was particular competition for such students from classes led by Herman Boerhaave, professor of physic and botany at Leiden, and Albrecht von Haller, professor of anatomy, botany, and surgery, in Göttingen.

Hill's solution to the lack of a suitable garden in London was to suggest that Kensington Palace could be developed for this purpose. Based on the organization of the Jardin du Roi in Paris, this would be funded from the royal purse, with free lectures on botany given on Saturdays when the king was absent.[51] Kensington, in Hill's plan, would be a worthy rival to the French botanic garden and become at once an ornamental garden suitable for a British monarch to enjoy as well as a scientific and educational enterprise—perhaps the ultimate example of the botanic garden, as it would be designed for both *utile* (utility) and *dulce* (beauty). Rather than develop a garden at Kensington, Hill instead obtained an unofficial position at Kew, under his patron the avid botanist Lord Bute, and produced the first catalog of plants being grown there by the Dowager Princess of

Wales in 1768.[52] Kew itself can perhaps be said to have finally developed into the type of scientific institution Hill imagined by the turn of the eighteenth century.

The Science of the Classroom

Up in Scotland, the Leith Walk garden in Edinburgh, created in 1763 by John Hope, fulfilled many of Hill's aims, albeit without a resident monarch and based considerably farther north. Hope was a member of the medical faculty at the University of Edinburgh, and his garden was designed with the explicit aim of teaching botany to an ever-increasing number of medical students.[53]

A novel physical feature of this botanic enterprise was the Botanic Cottage, which was built into the external wall of the garden. This was used as both a lecture room (upstairs, fig. 1.5) and residence for the head gardener and his family (downstairs). The garden itself was an essential teaching tool. As Noltie has argued, Leith Walk represented a new type of permanent teaching and research institute inspired by examples such as the Jardin du Roi.[54] Given that Hope visited several European centers of botany before successfully petitioning for a garden in Edinburgh, such continental stimuli must have influenced his design. This continental experience, particularly studying under the eminent French naturalist Bernard de Jussieu, may also explain the importance that Hope placed on botany as a subject in its own right.

Like Boerhaave's garden at Leiden, which O'Malley has described as being laid out with geometric regularity to "underline the order of Boerhaave's abstract system" of ordering plants, the design of Hope's garden also appears to have reflected his own particular scientific focus on the creation and dissemination of botanical knowledge.[55] At Leith Walk the traditional medicinal plant beds of regimented straight lines used for teaching only formed a small portion of the garden scheme, known as the "Schola Botanica," and were relegated to one side of the plot. The core design in front of the main glass houses was more elaborate, and from extant plans, such as the 1777 delineation, appears to have incorporated an organic lay-

Fig. 1.5. The lecture room, located on the floor above what would have been the gardener's living quarters of the recently relocated and lovingly recreated botanic cottage. Now used by community groups and as a space for public events. Reproduced with permission of the Trustees of the Royal Botanic Garden Edinburgh.

out with a series of paths shaped around beds as if following the outline of a poppy seed head (see plate 3). This design can also be seen to be an example of contemporary landscape gardening tastes, with their informal beds and winding walks.[56]

This was very different from the more formal designs found in the earlier university botanic gardens. Although, as O'Malley has outlined, even the design of these apparently rigid gardens went beyond strictly instructive purposes, as is clear from the complexity of their plans.[57] She has also noted that by the eighteenth century botanic gardens were being adapted in order to place plants in artificial reconstructions of the natural environments in which they were discovered, and she suggests that "contemporaries believed that the observation of plants in a naturalistic landscape that imitated original habitats or gave some sense of their natural affinities would better serve the scientific and educational function of botanic gardening."[58] It is likely that these changes in the idea of how plants should be displayed, as well as shifting conceptions of domestic garden design, also played a role in Hope's more stylistically informal design.[59]

This change also reflected a growing interest in botany as a scientific system following the publication of Linnaeus's *Philosophia Botanica* in 1751 (and the nurseryman James Lee's English summary of the text, which was reprinted eight times between 1760 and 1811).[60] Hope used this system to lay out elements of the garden and even installed a monument to Linnaeus in the Leith Walk garden in 1779.[61] Linnaeus himself also highlighted the economic advantages that could potentially accrue from understanding this new botanic knowledge, which made it of even greater popular interest.[62]

Hope's lecture notes themselves highlight that the predominant focus of his course was botanical science, and Hugo Arnot, in his 1789 *History of Edinburgh*, remarked of Hope:

> In the first part of his course, he treats of vegetation, several parts of which he explains by a variety of experiments in the Botanic Garden. In the second he unfolds the botanical system, and treats fully of the natural order of plants. The third is devoted to the explanation of the nature and use of exotic plants, the whole being concluded with a history of botany . . . and the students in general, have freer access to the garden, than is permitted in foreign universities.[63]

This evidence is in line with O'Malley's argument that in this period botanic gardens became "museums of living plants," and were transformed into "centers for research experiment, display and delight."[64] It is also clear that as well as being experimental spaces, part of their particular appeal was the amount of access given to students to the garden spaces. In particular, this is seen as more extensive than in other continental European examples, which underscores the important role botanic gardens played in the competitive student marketplace.

The central role of the garden and the physical plant material for use in teaching can also be seen in the student notes. As already noted, in the third part of the lecture course, Hope is recorded as stating that he "will hold up to view the different things in the Garden."[65] So the garden provided material that could be brought into the classroom for teaching purposes as well as a place where plant specimens could be viewed closely. This approach also mirrored that of Boerhaave in Leiden, where "in summertime, at seven o'clock in the morning, Boerhaave surrounded by his students, could be seen in the Garden, lecturing on botany and demonstrating the plants."[66] The influence of Boerhaave was particularly strong in Edinburgh as, from the 1720s until the end of the century, the medical faculty modeled their approach to teaching on his as well as utilized his texts.[67] Hope's predecessor and teacher Charles Alston had also trained in Leiden and exchanged plants and seeds with Boerhaave over a number of years when supervising the earlier university botanic garden, so this would no doubt have also influenced Hope when he was establishing the Leith Walk teaching space.[68]

However, the use of the garden for teaching did come with problems as the student body taking the course grew. According to the student notes, Hope described how the third

> & last part of the course consists in demonstration, in this part I have made some improvement & in others I have given it up entirely, for a no. walking thro' the Garden, especially in bad weather I found to be of great injury to it. It is impossible for you to have access to the Exotic plants but by demonstration yet many of the Exotics do not come to such perfection as to admit of demonstration.[69]

So the plants, if left in situ, could be damaged by the ever-increasing number of students trying to view them, and even attempts to bring specimens indoors to protect the plants could be thwarted by their inability to acclimatize and produce successful blooms in the artificial conditions of the greenhouse.

Gardeners were essential to both the maintenance of the plants and this approach to botanic teaching, which placed the garden at its center. At

Edinburgh, the first head gardener to live in the botanic cottage was John Williamson. He was head gardener for twenty years, from 1760, through the establishment of the Leith Walk garden, until his death in 1780. His personal role within the Leith operation can also be found as a trace in the early physical layout of the garden. Not only was the cottage marked as Mr. Williamson's house on the 1777 plan, but one of the plant beds in front of the greenhouse was also labeled in Williamson's honor. Having a named bed placed Williamson in exulted botanical company. Other beds in this section of the garden were named after prominent botanists, both contemporary and historical, as well as patrons, including the influential Bute, and others who exchanged botanical specimens within Hope's extensive and powerful network.[70]

After Williamson's untimely death in 1780, Hope installed a memorial to him in the garden, which recorded that he was "esteemed for eminent skill in his profession."[71] That, along with the naming of the bed, suggests that Hope respected his horticultural expertise. This type of monument set within a garden was generally only reserved for more elite figures. At Leith Walk the only other memorial was the urn dedicated to Linnaeus. The memorial to Williamson is a physical reminder of the interlinking network of the garden, gardener, and professor. As Stephen Harris notes in relation to Oxford, the greatest periods of success for the garden were when "relationships between horticultural and academic staff are strongest."[72] There is no reason to suggest that this would have been any different at Edinburgh.

At Leith Walk, the purpose-built cottage, with lecture theater above the domestic rooms, emphasized the important role of the botanic garden as a teaching space, and the gardener's place within that sphere. As we have seen, Hope's main focus was on the teaching of botanical science rather than the medicinal uses of plants. At Edinburgh, as well as needing someone to both grow and demonstrate plants for the students, Hope also employed Williamson to conduct experiments to accompany his lectures. In the lecture notes, Hope explained that in the Leith Walk garden "we are making experiments here, but experiments on trees and plants are very different from those on animals."[73] These experimental examples formed the core of much of his teaching, and Hope referred to them throughout his lectures. This suggests that the gardener showed the actual experiments to the students or, where this was not possible, specially commissioned teaching diagrams depicting the results.

Williamson kept a notebook outlining the experiments he was conducting in the garden and their results, as well as a list of experiments that were physically based in the gardener's house made by Hope.[74] From this, we can see that Williamson was acting as an invisible technician in common with other areas of scientific endeavor.[75] The experiment notebook seems to contain two sets of handwriting, which perhaps indicates the co-production of knowledge between Hope and Williamson and again blurs the division between head and hand. In line with Steven Shapin's argument of the dual invisibility of technicians in the past as well as to historians, it is clear that Williamson's research was appropriated by Hope without credit. For example, Williamson undertook hybridization experiments between the oriental and opium poppies, but when Hope reported the experiments to the younger Linnaeus, he omitted Williamson's role in the process.[76] This perhaps indicates that there were limits to the role a skilled technician or gardener could play within the hierarchy of the botanic garden.

Gardeners were an integral part of the teaching process at Edinburgh in other ways more befitting their social status. The fee for the botany course, which was collected by Hope, was two guineas for each of the summer sessions, rising to three guineas in 1770.[77] The collection of these fees was used in part to cover "payments to the gardeners for assistance with setting up, running and tidying up after the lectures," which establishes the important role gardeners played as facilitators for the delivery of botanical lectures.[78] Further evidence of the practical roles gardeners played in Hope's classroom can be seen in his own notes. In 1778 he recorded that "one of the gardeners should keep a register of the students examined."[79] He mentions gardeners holding up specimens of "waking" plants for students to compare with a drawing of them "sleeping."[80] This reflected a widespread interest in plants that moved their leaves at this time, but it also indicates the use of a variety of pedagogical tools and the role drawing played as a way of knowing, which will be explored in more depth later.

"Because a certain number of different plants, all in flower, must be had for each lecture"

While Edinburgh gives us an insight into Enlightenment botany teaching using a well-stocked garden, on-site classroom space, and a team of gardeners acting as scientific technicians, the tale of Glasgow offers us an insight into the problems faced when these resources are severely limited. Like Edinburgh, the University of Glasgow was teaching botany to its medical students. However, since the 1740s, the Glasgow medical faculty staff had been complaining about the state of their existing physic garden. In 1754, the eminent physicians Cullen and Robert Hamilton delivered a proposal to the faculty in which they argued that the garden needed major improvements in order "to make it more useful to the study of Botany."[81]

As well as bemoaning the current state of decay of the fruit trees, the nature of the soil, and its "situation very much exposed to the smoke and the soot of the town," they argued that the university should "take the proper measures for planting their Garden in a manner becoming a Society devoted to Taste and Science."[82] They concluded by stating that "on this occasion we cannot avoid observing that the study of Botany in this University has been very much retarded by the want of a proper Gardiner & that the present appointments are insufficient for engaging one."[83] This request had no discernable impact, but it is clear that a good gardener and a decent garden were essential for those involved in the delivery of botanic teaching by the mid-eighteenth century, particularly when teaching took a sensory approach and required the expert handling of a large array of specimens by each student.

In 1806, Thomas Brown wrote to Dr. Jeffray, joint professor of anatomy and botany, whose botanical course Brown was teaching on his behalf, in response to criticism that had been leveled at the gardener William Lang, regarding the management of the old botanic garden. He wrote:

> I am very sorry to learn that the College is dissatisfied with William Lang's behaviour & I am much afraid that it has been improper in many respects, but I can only say that as far as

> the Botanical Department is concerned I have no fault to find, but every reason to be completely pleased with it. That plot of ground which is dignified with the name of Botanic Garden is so very barren, that its produce can scarcely be of any advantage to a lecturer of Botany. He is therefore, under the necessity, during the greatest part of the course, both of collecting plants himself in the fields & in neighbouring gardens, & of trusting to the exertions of the gardener.[84]

From this we can see that similar issues to those outlined by Cullen and Hamilton had continued regarding the state of the botanic garden. However, by now it had been reduced to such a poor state that Brown in the same letter stated that "I have so little to shew the students, everything looks so meager, that I was even doubtful of the propriety of raising the fee."[85] This lack of a decent botanic resource also had a further indirect effect on the tasks conducted by the gardener.

On January 25, 1807, Lang himself wrote a detailed representation to the committee in response to the complaints made regarding his ability to fulfill his duties. Given its importance as a firsthand account from Lang describing the method of collecting specimens and their use for teaching, it seems important to include this quote at length.

> During the summer when the Botanical lectures are going on, the garden allotted for that Department furnishing but a very few specimens for illustrating the Science of Botany, it is required of me to collect elsewhere whatever plants may be necessary for carrying forward the lectures. For which purpose I have to traverse the country round in search of plants: and that, Gentlemen, not on a particular occasion but almost every day of the course. A great part of my time therefore which should be devoted to dressing the Gardens is occupyed in this manner. Because a certain number of different plants, all in flower, must be had for each lecture. And oftentimes after, I have travelled to a wood or waterside two or three miles from Town. I have been

> disappointed in finding the individual plants wanted—and must again set out to some other quarter to find them. And Gentlemen, as the number of students last year was upwards of thirty, it became necessary for me to provide upwards of thirty specimens of each individual plant demonstrated. And as several hundred Genera and Species were examined last season, the Botany Garden not furnishing near one hundred in perfect condition. A great proportion of my time must be occupied in this manner.[86]

Here we get a detailed firsthand insight into the way specimens were used in the classroom, the role of the gardener as a botanical assistant or technician, and the problems encountered when a decent botanic garden was not available in close proximity to the classroom. It also raises interesting questions about where botanical collections were located in this period and how material was moved from other gardens and the field to the classroom. In doing so, it also transforms the Rousseauian vision of the gentleman botanizing at leisure (as depicted in fig. 1.1) into a difficult, time-consuming, and potentially costly act.[87]

Sadly for Lang, neither his testimonial, nor that from Brown, appears to have been enough to convince the committee that he was a proper gardener, and for whatever reason by the end of 1807 he was no longer employed by the college. Nothing else seems to be known about him after this time. Apart from Brown, who relied on Lang to supply his students with the necessary resources in the classroom, the faculty appear not to have understood his integral role as an essential technician in the delivery of the new botanic lectures, with their requirement of a range of specimens for each student so that they could observe, dissect, draw, and thereby understand their subjects using their own senses.

Knowing through Sensory Engagement

As the description of specimen collecting by Lang suggests, the role of objects, whether living, preserved, or recreated in model form, was of crucial importance for teaching at this time. As Easterby-Smith writes, "Learning and practicing botany in the eighteenth century involved collecting specimens, identifying them and conserving them (or their representations) for future reference."[88] The nature of botanical knowledge creation therefore included live specimens in gardens, dried examples in herbaria, and illustrations and descriptions in books and letters. This meant that at the center of botanical and other natural history collection building was both a gathering together and an exchange of knowledge and objects, which also made them particularly sociable disciplines.[89]

There were parallels too with the delivery of the teaching of anatomy, with its use of specimens as key learning tools (plate 4). In 1784, William Hunter, a leading anatomist and physician, described how small specimens were to be passed out around the room, one student describing to the next what was to be seen.[90] There are no clear descriptions of what the students did with the botanical specimens at Leith Walk or in Glasgow, but we can speculate that the pedagogical approach would have been similar with the more exotic, and thereby rarer, productions of nature. From Lang's account it seems that more common specimens would have been provided in a great enough number for students to have their own specimens for dissection and close observation.

This relationship between the sensory teaching of anatomy and botany can be examined through the multifarious roles of Andrew Fyfe as an artist, botanist, and anatomist. His story illustrates how the skills learned within the new Enlightenment botanic garden could be appropriated and applied directly to medical practice, in a way that is missed by historians as well as the author of the *Guide.*[91] Fyfe first appears in the record as a gardener in the Leith Walk garden between 1772 and 1775, and there is a note mentioning him as collecting plants while accompanying Hope in 1773.[92] However, at the same time he was also an artist and attended drawing classes at the Trustees Academy in Edinburgh in 1760, and in 1776 he was reported as being the winner of a prize for his drawings of flowers and foliage.[93] The same news article that reported this win also stated that he was no longer an assistant gardener but was now a student of physic.[94] We can only speculate, but as suggested by Noltie, Hope probably facilitated Fyfe's move from assistant gardener to medical student.[95]

Fig. 1.6. One of Andrew Fyfe's illustrations used to accompany Hope's lectures. It depicts an experiment repeated at Leith Walk and originally based on a description in Stephen Hale's 1727 work *Vegetable Staticks*. The note in the top left corner records that "This is shewn at the lecture on the motion of the sap." Reproduced with permission of the Trustees of the Royal Botanic Garden Edinburgh.

In 1776 Hope wrote a note regarding improvements to be made to his teaching practice, in which he states that "every figure necessary to be shewn drawn by Andrew Fife and of such size that it may be seen at any distance in the room, a note of these drawings made out."[96] This suggests that Hope was promoting Fyfe's skills as a technical artist and, although he was a student, employing him to create large botanical drawings as teaching aids for the classroom while he was training in medicine (fig. 1.6). This role is muddied by the fact that while acting as a student, a gardener, and an illustrator, from 1771 he was also the principal janitor and macer in the university (a post then usually held by a student, as it provided free living quarters), and, from 1777, a dissector for the anatomy department.[97] Fyfe remained in this multipurpose post for forty years and became a renowned demonstrator of anatomy.

The *Guide* records:

> For the benefit of those who wish to acquire a perfect knowledge of anatomy, private demonstrations of the subjects of Dr Monro's lectures on the structure of the body are given at an evening hour by Mr Fyfe. Every gentleman should attend this course the first season he attends Dr Monro, for by doing so, he will see every part more distinctly than the crowd at the doctor's class will allow and besides he is entitled to witness the preparation of the various parts of the dead body, which is necessary for illustrating the lectures.[98]

Here, the body was demonstrated in similar ways to the plants and used to illustrate the lectures. It suggests that sensory ways of knowing, developed as a gardener working with plant physiologies and as an artist, could also be applied to understanding the human body.

A central approach to Fyfe's understanding of the material of nature in all its forms was his use of drawing. As well as drawing plants for Hope, he also produced anatomical textbooks that included his own illustrations. Thus, the approach of learning through drawing, the training of hand and eye, might go some way to explaining his skilled draftsmanship, as learning through illustration was part of botanical, and perhaps also horticultural, training.

This use of an artistic approach to knowledge creation has clear parallels with Fyfe's contemporaries, the anatomist and artist Charles Bell and Joseph Black, who lectured on chemistry at Edinburgh, since they both used illustrations in their pedagogical technique.[99] The parallels between Fyfe and Bell go deeper, given that there was also a shared use of other media within the classroom. As Carin Berkowitz has described, Bell's pedagogical approach also included "wax models; preserved specimens in jars, housed in collections; schematic chalk drawings from the classroom; elaborate engravings and less elaborate etchings found in books; dead bodies; paintings and sculptures and living bodies."[100] This interrelated set of tools was mirrored by the variety of materials used in teaching botany. A list made by Hope related to his teaching activity in 1776 included "specimens of dried plants which were pasted to the paper; after finishing lecture I showed 15 to 20 more which were loose in the paper"; some "dried in sand and preserved in glass"; the illustrations he desired to be made by Fyfe and within which he also noted that "what can be put into such a form as to be handed about should be done."[101] These different objects, texts, and live specimens were used for both research and teaching purposes and often worked together, as Berkowitz defines, "in a science that was rooted in the classroom."[102] So the scientific space of the botanic garden can be seen as both creating and disseminating knowledge through these varied materials.

These shared research and pedagogical approaches using a range of material elements may have led to the development of other shared resources outside of the confines of particular spaces, such as the botanic garden or the anatomy lecture theater. In the Glasgow faculty minutes there is a draft list of regulations for the use of the Hunterian collection. Within this list the minutes state that "the Professors & Lecturers of Anatomy, Botany, Natural History, Midwifery & Materia Medica shall have access to the Corresponding Departments of the Museum, and the privilege of borrowing from it such preparations, specimens and articles as may be necessary and useful in their Public Lectures & Studies."[103] Perhaps then we also need to consider the way subjects were taught based on the material objects used and who else, apart from the professor or other lecturing staff, was involved in their creation and use. It also begs the question of how far both concepts and materials were shared between different disciplines within the university. Places such as the botanic garden and the anatomy or chemistry lecture theater are generally viewed as discrete silos, but as we have seen, both academic and technical staff as well as students moved between them, so there must also have been movements in pedagogical ideas and possibly objects, too.

Fyfe also acted as the curator of a collection of anatomical figures given to the university by Alexander Monro (Secundus) in 1800, which suggests that he was interested in the material culture of teaching as well as that of illustration. Berkowitz discusses the use of collections of objects in Bell's anatomy classes at the Great Windmill School in the early nineteenth century. According to her, "those objects helped to constitute a pedagogical program at the center of Bell's medical science, in which surgery and general medical practice were taught through the cultivation of sensory perception and the training of hand and eye, such that accumulated sensory experience could be, at more advanced stages, generalized and systematized."[104] As we have already seen above, Hope's aim was to encourage a sensory understanding of botany with his encouragement to students to develop their sensory perception so there was a shared understanding of the role of sensory experience in both botanic and anatomical knowledge creation.

The botanic garden then was a teaching and research laboratory in which students were trained to use their senses in order to understand the natural world. Rather than being an isolated Edenlike world, it was a vibrant center of expertise that shared pedagogical tools and sometimes even personnel with other scientific areas of study, such as anatomy. By viewing the garden as a multisensory and multimedia teaching space, it becomes, for us, more than a collection of plants and takes on new meaning as an Enlightenment space where the world was being explored, classified, and systematized through a variety of methods. Such attempts to understand plants were not confined to university medical schools and their students, but led to the creation of new types of botanic gardens developed by private collectors, which attracted scientific scholars as well as a public excited by the opportunity to view the latest exotic imports.

TWO

Creating a Perpetual Spring

Tracing Private Botanic Collectors and Their Networks

THE INTEREST IN AND SOCIABLE attachment to botanical knowledge, which were encouraged by a medical education, meant that professional men with a reasonable disposable income and a passion for botany sought to develop and demonstrate this skill in their own private gardens. Lettsom describes his garden at Grove Hill (plate 5):

> The lower extremity opens into the Arbustum, through which a walk of nearly a mile in extent is carried under the shade of upwards of one hundred fruit-trees, which not only form a pleasing shade, but likewise prove objects of beauty in their blossoms, and of profit in their product. On the borders of this walk grow about four hundred European plants, placed in succession agreeably to the Linnaean classification, and lettered in legible characters, a catalogue of which is preserved.[1]

In this excerpt from his guidebook, Lettsom portrayed his domestic garden as a place full of trees and plants that he considered to be both ornamental and productive, as well as a section organized scientifically using the Linnaean system of classification.[2] Pettigrew, as always a useful guide to Lettsom and his landscape, suggested that "any person, however ignorant of practical botany, might acquire a tolerably correct idea of that valuable science, by a due attention to the arrangements, &c." of the garden at Grove Hill.[3] This arranging and labeling of plants establishes that scientifically arranged botanic collections were not only confined to university-owned spaces, but that they were also created and developed by individuals. The role of guidebooks and labels in transmitting and codifying this knowledge for others will be explored in detail in chapter 4, but it is clear that the garden was designed as an educational and experimental space as well as a place for personal leisure.

Gardens such as Grove Hill formed a crucial node within larger networks between which plants, people, and knowledge circulated. Such places were of key importance for the imperial project, acting as colonial botanic laboratories in which activities that were integral to the needs of the empire were performed.[4] Londa Schiebinger argues that key to this were the royal and imperial botanic gardens that acted as experimental stations, both in terms of the acclimatization of plants so that they could be grown in various regions of the world, as well as places in which economically valuable plants could be trialed and understood.[5] We can expand this group of nationally strategic institutions to include the private botanic gardens established by men such as Lettsom.[6] This broader network acknowledges that within this "imperial geography of plants" there were many gardens of all types and sizes as well as numerous gardeners of all levels.[7] Miles Ogborn, for example, notes that

> alongside the Bath botanical garden, and often growing the same plants, were the extensive private gardens of wealthy gentlemen such as Matthew Wallen and Hinton East; the plantation gardens of horticulturalist slaveholders and overseers; and the provision grounds of the enslaved themselves. The latter have been called the "botanical gardens of the dispossessed" and were, despite the claims of other gardeners, where vital food crops were nurtured, many with African origins.[8]

All these gardens at home and abroad formed part of the imperial horticultural network with Britain's keystones of national gardens, which included Kew in London as well as botanic gardens established in Saint Vincent in the West Indies, Jamaica, and Calcutta. These key imperial spaces were the foci for scientific work on plants brought in from around the globe, as well as providing the locus for the transfer of plants to other gardens and fields. This approach resulted in new economically important crops for the various colonies and, as Louise Brockway argues, "thereby altering the patterns of world trade and increasing the plant energy, and human energy in the form of underpaid labor, that the European core extracted from the tropical peripheries of the world system."[9] Private gardens like Lettsom's can be seen to play a significant, although less official or visible, role in these networks.

Correspondingly, an interest in popular science such as botany allowed networks to flourish and develop. Shared concerns, such as botanical knowledge, were as important as the profession or position of those people involved.[10] These connections could then involve patronage of those conducting activities, such as plant collecting, as well as creating ties with those of higher status, such as the landed gentry, through the circulation of seeds and plants. Private gardens, which we can also extend to include semipublic gardens and commercial nurseries, were part of wider botanical networks between which plant material, objects, knowledge, and people circulated alongside the more visible named botanic institutions.[11] One early example of this, discussed by Esther Arens, illustrates the type of circulation within the wider European network. Gaspar Fagel, an adviser to William of Orange, developed his own botanic collection at his estate in Leeuwenhorst in the Netherlands.[12] After his death, part of his plant collection was transferred to Hampton Court in 1689 and placed in a new "glass garden." Not only was the plant collection moved to a royal garden, but it was accompanied by expert Dutch gardeners who were able to document the plants as well as expand the collection.[13] This movement or circulation between botanic collections again underlines the important role of private gardens with larger networks and emphasizes the movement of knowledge through both the transfer of plants themselves and their accompanying expert gardeners.

This categorization of private gardens as distinct from university botanical spaces means that they have rarely been studied as locations of knowledge creation, as they are predominately viewed as spaces for pleasure and leisure.[14] However, as we have already seen from Lettsom's example, taxonomical and scientific collections could also be maintained on

in the hot house."[21] The collecting of plants and the spotting of an exotic live chameleon are not activities that you would normally assume to have occurred within a garden designed purely for botanical and educational purposes.

On the same trip to London she writes that she had traveled again with her niece, Georgina, and also Mrs. Port, "to Upton in Essex, 10 miles off, to Dr Fothergill's garden, crammed my tin box with exoticks, overpowered with such variety I knew not what to chuse! Georgina delighted fluttered about like a newborn butterfly, first trying her wings, and then examining and enjoying all the flowers."[22] Here the doctor's garden provided an exciting range of exotic introductions, which were both of scientific and pleasurable interest and from which specimens could be gathered and taken away. Whereas a trip to Lee and Kennedy's popular plant nursery, The Vineyard in Hammersmith, was described as only offering "a pleasant tour this morning."[23] She continues: "We went to Lee's at Hammersmith in search of flowers, but only met with a crinum, a sort of Pancratium Crinum Asiaticum."[24] Obviously she was disappointed by seeing a sole example of a flowering plant that she had not seen before. It is worth noting here that Mrs. Delany was not only an elite woman, she also had a very specific botanical interest. As an artist she developed an incredibly skillful approach to producing paper collages known as "mosaicks," which were botanically accurate as well as beautiful, and this was no doubt related to her particular approach to seeking out new and exotic flowers.[25] However, this fluidity of use of botanic collections, whether educational, private, or commercial, reflected the blurred nature of what constituted a botanic collection at this time and how they were experienced by visitors.

This circulation of plant material between these spaces and the wide-ranging networks this encompassed is also made visible in the 1778 *Proposals for Opening by Subscription a Botanic Garden to be Called the London Botanic Garden,* written by William Curtis, an apothecary by training.[26] In this text, Curtis gives thanks to those who have supplied him with plants for his new botanic garden, which he states that he plans to fund via individual subscriptions—a membership scheme by which visitors paid a regular sum in order to access the garden and its library. In his record of thanks to those who had helped establish the garden are references to specimens donated from King George III's royal garden at Kew, as well as the private gardens of the Earl of Bute, the Duchess Dowager of Portland, Dr. Fothergill, and Dr. Pitcairn. He also thanks the Apothecaries Company, which we can assume relates to donations from their physic garden in Chelsea, as well as a set of London-based nurserymen: "Messrs Gordon, Lee, Kennedy and Malcolm."[27] All of this again demonstrates the existence of a network that crossed private, institutional, and commercial botanical collections and one that was connected by the movement of objects, plants, seeds, knowledge, and people. It was also one in which medical practitioners played a prominent role, as we can see from Curtis's list and which will be discussed in more detail in chapter 3.

The interconnections and movements between gardens are also illuminated by a close analysis of the gardens of Fothergill and Pitcairn. Like Lettsom, both men were key medical figures at the end of the eighteenth century, with large disposable incomes and estates on the fringes of the city, as well as busy central London medical practices. Together they funded plant-hunting expeditions and used their gardens as botanical clearing-houses as well as training grounds for apprentice gardeners.

Throughout Britain in this period, medical physicians developed gardens in which seeds and plants could be grown, observed, and circulated. In this way they formed part of larger gentry and university botanic networks, and thereby operated in similar ways to the more commercial gardens of nurserymen.[28] As we have seen, private gardens with botanic collections were not unusual during this period. Other examples include the Duchess of Portland, who had "every English plant in a separate garden by themselves" at Bulstrode, and there were specially created botanical and experimental gardens at Woburn Abbey, Buckinghamshire, developed by the 6th Duke of Bedford, who was an important patron of scientific horticulture.[29]

This network of private gardens also extended beyond British shores. In 1768 Fothergill wrote to John Bartram, an American plant collector and exporter, exhorting him "to sow a considerable part of most of the seed thou collects, I mean the new discovered plants, in a little garden at home,

Fig. 2.2. Representation of the house on Little Jost Van Dyke in the British Virgin Islands, where Lettsom was born in a Quaker settlement in 1744, as illustrated in Pettigrew's *Memoirs of the Life and Writings of John Coakley Lettsom,* 1817. Wellcome Collection, CC BY 4.0.

and to send over young plants of two years old in boxes, several sorts of plants in one box."[30] His argument was that the young plants would survive better as "many seeds wholly miscarry with us," which demonstrates the importance of these domestic private spaces as places for growing plants within such networks.[31] In this way the private garden could act as a nursery space from which plants could make their way to other private, subscription, commercial, and institutional gardens, either at home or abroad.

Although this book is focused on the gardens created in Britain, it is important to consider that this circulation, whether of people, plants, and/or objects, was intertwined with histories of slavery and domination over indigenous people and cultures. Lettsom himself was born on Little Jost Van Dyke near Tortola, one of the British Virgin Islands, which housed a Quaker colony as well as a number of slave plantations (fig. 2.2). Pettigrew tells us that in the 1760s, Lettsom, having been educated in Britain, returned to his native island to take possession of the property which had been left to him by his father and "which then consisted of a small portion of land, and about 50 slaves."[32] Pettigrew continues that although at this time he did not have as much as fifty pounds sterling to his name, Lettsom considered "the traffic in living blood as wicked and unlawful," so that "he immediately *emancipated* them and became a voluntary beggar at the age of 23."[33] Despite such a notable early attempt at the emancipation of enslaved people, Lettsom inherited another similar plantation shortly before his death and without time to achieve a second emancipation, demonstrating the interlinking of eighteenth-century lives, whether willing and intentional or not, with the more violent elements of empire.

Similarly, the seemingly benign traffic in plants from remote parts of the world was necessarily tied to the trade in people, since ships were used for more than one purpose as they crisscrossed the oceans. Kathleen Murphy, in her groundbreaking work, has traced the direct relationships between the apothecary and naturalist James Petiver's natural history collection and the trade in human cargo across the British Atlantic.[34] As she notes, "while the scale of Petiver's efforts was extraordinary, his use of the global routes of British commerce to expand his collections was not."[35] This common use of the same global routes would also be true of the medical practitioners discussed. They may not have been directly involved in slavery—or in Lettsom's case may have attempted to free themselves from profiting from human misery—but they all still relied on these trade routes and connections to create their collections of new and exotic plants. As James Delbourgo has argued, "Only in the last few years have scholars begun to examine the agency of the slave trade in circulating natural knowledge, suggesting the possibility of overcoming the long-standing notion that slavery and science had nothing to do with each other."[36] This relationship between slavery and science is particularly embedded in the natural history collecting networks of medical practitioners and the crucial role of ships' surgeons, who often had extensive botanical, zoological, and medical knowledge. Almost half of the maritime men, which included a large number of surgeons, who were collecting for naturalists such as James Petiver in the Atlantic were doing so along slave trade routes.[37]

Given the focus on the experience and use of gardens by medical practitioners in Britain, I have not attempted to trace here the routes of plants that made their way into the various gardens, but it is clear that this history of collecting exotic species is entangled with other forms of commerce as plants traveled on the same ships that transported other cargo across the empire. There is further work needed to be done on the context of the collecting and distribution of plants in order to understand the full economic and human cost in the creation of our historic landscapes, although Tobin's work has already made crystal clear the necessity of understanding the other histories that have led to the creation and cultural shared understandings of the exotic and its placement in our gardens.[38] Underlying all these narratives of national benefit, of sending people to Africa to collect plants and trade connections with the Americas, is a darker history of violence and exploitation. This necessarily means that the mention of the "exotic" comes freighted with hidden histories of labor, both abroad and at home, which are sometimes hard to identify from the extant sources.

"A Sensual Botanist"

Taking this circulation of natural history and botanic specimens as our starting point, it is clear that these gardens were all constructed via networks on a range of scales from the global to the truly personal. Moving from the global frame to the individual, it is worth also considering the personal connections that impacted on the movement of plants across garden spaces. Having regularly frequented the estate of Dr. Fothergill (fig. 2.3) at Upton while the older physician was alive, Lettsom obtained two thousand botanical specimens, a fraction of those growing in the garden, along with their attendant greenhouses, transferring them to Grove Hill on Fothergill's death in 1780.[39] The movement of plants from one collection to another illustrates their importance both scientifically as well as emotionally, as they represented a memorial to Lettsom's mentor, friend, and key Quaker connection.

As a Quaker with a degree from Edinburgh, at a time when the Royal College of Physicians only licensed those with a degree from Oxford or Cambridge, Fothergill's career as an unlicensed doctor in London got off to a shaky start, with him barely able to make ends meet. However, in 1744 his

Fig. 2.3. Portrait of John Fothergill with a botanical text in his hands and sitting on a chair covered in a pineapple print, denoting his deep interest in exotic plants and botanical science and highlighting the ever-present colonial context in domestic settings of the time. Mezzotint by V. Green, 1781, after G. Stuart. Wellcome Collection, CC BY 4.0.

fortunes changed, and after being examined by the Royal College, he became the first graduate in medicine from Edinburgh to be granted a license to practice. From that point onward his career developed along a rapid upward trajectory until he became one of the richest physicians in England.[40] This was in part due to the success of his treatise, *An Account of the Sore-Throat Attended with Ulcers,* published in 1748.[41]

His Edinburgh connections remained important and included his lecturers, such as the surgeon and anatomist Alexander Monro, and Charles Alston, professor of botany and materia medica (who was John Hope's predecessor), as well as contemporaries such as the surgeon William Hunter. From later letters it is clear that Fothergill and Alston retained a botanic friendship long after he graduated.[42] As well as describing his time as a young physician trying to set up business in London, Fothergill also filled his letters to Alston with whatever botanical knowledge was being discussed and circulated at the time.[43] This suggests his botanical interest was fostered by his time as a medical student at Edinburgh and grew as his fortune allowed.

Other members of his network, including fellow Quaker and banker David Barclay, may also have provided encouragement for this botanical interest. It was Barclay who introduced Fothergill to Peter Collinson, a well-off Quaker merchant who traded mainly with the American colonies and the West Indies, and in 1740 Fothergill recorded his pleasure in this new friendship.[44] Later in 1774, Fothergill wrote to Linnaeus that "it was our Collinson who taught me to love plants. . . . He persuaded me to create a garden."[45] Collinson himself was at the center of an international network of naturalists and botanists, and his own gardens, firstly in Peckham and then at Mill Hill in Hendon (both on the rural fringes of London), were used as growing grounds for new plants from around the world.[46] This was a particularly important link for Fothergill, and brought him into contact with others, such as the American plant collector John Bartram, as well as Linnaeus. It is through this network that Fothergill became involved in plant collecting, particularly via Bartram in America, and his house and garden in Upton also became a repository for plants, animals, and a collection of shells, corals, and insects.[47]

In 1762 Fothergill purchased Admiral Elliot's estate in Essex and established his main garden there (fig. 2.4). Known as Upton House (now the public West Ham Park), the most extensive estimate, which is from Gilbert Thompson in 1782 just after Fothergill's death in 1780, recorded it as "containing about sixty acres of land, and between five and six acres of garden-ground."[48] Given that Lettsom stated that the original estate when Fothergill purchased it was estimated at thirty acres, this larger estate prob-

Fig. 2.4. John Fothergill's garden at Upton, and later the birthplace of pioneering surgeon Joseph Lister, which depicts the essential team of gardeners as well as exotic plants and birds. Engraving as reproduced in A. Logan Turner, *Joseph, Baron Lister: Centenary Volume, 1827–1927* (Edinburgh & London: Oliver & Boyd, 1927). Wellcome Collection, CC BY 4.0.

ably included the further parcels of land Fothergill bought as he developed his landscape.[49] These additional parcels were used for various tree plantations, including Portuguese oaks and Spanish chestnuts, emphasizing the arboricultural and possibly agricultural nature of the estate, as well as its role as botanic garden.[50]

However, given the exotic nature of much of the plant material grown at Upton, it is not surprising that the botanic infrastructure included both hot and cold greenhouses. Lettsom, presumably based on his many visits to Fothergill's garden, describes how they were of nearly twenty-five feet extent and that they communicated directly with the house via a glass door.[51] Within these were "upwards of 3,400 distinct species of exotics," with around another "3,000 distinct species of plants and shrubs" growing in the garden outside.[52] Lettsom portrayed this delightful scene as a "perpetual spring . . . where the elegant proprietor sometimes retired for a few hours, to contemplate the vegetable productions of the four quarters of the globe united within his domain; where the spheres seemed transposed, and the arctic circle to be joined to the equator."[53] This Edenic flattening of the globe within the garden was both a symbolic organization similar to that discussed in the last chapter in relation to the earlier Padua botanic garden, as well as a physical manifestation of the vegetable productions of the British Empire, which was stretching ever farther across the known world. It is, therefore, perhaps unsurprising that Banks declared that "no other garden in Europe, royal, or of a subject, had nearly so many scarce and valuable plants."[54]

In many ways this collecting seems in line with that taking place at other institutional gardens and at Kew. On Fothergill's death, Banks and Solander described in a note how "the remembrance of his botanick garden at Upton will ever be fresh in the minds of all lovers of that science."[55] However, it is clear that for Fothergill his botanic collecting had more personal significance beyond that of a collection based on scientific rigor. In 1772 he wrote to William Bartram, who was collecting plants for him in South Carolina: "All fragrant shrubs or plants, or such as are remarkable for the beauty or singularity of their flowers and foliage will be most acceptable. I am not so far a systematic Botanist as to wish to have in my garden all the grasses or other less observable humble plants that nature produces. The useful, the beautiful, the singular or the fragrant are to us the most material."[56] He echoed this theme again in 1774 when he wrote to Lionel Chalmers, declaring "I call myself a sensual botanist."[57] This sensual approach, he explained, is the reason why only "plants remarkable for their form, foliage, elegant flowers, utility, are my objects. Mosses, grasses and the like I leave to others. Ferns indeed and the *Polypodiae,* I love. They are all elegant."[58] This is a man then who was not just collecting plants by scientific principles but also choosing them based on personal, sensory criteria. They had to be either of a pleasurable nature to be enjoyed as such or be of utilitarian value. This description acts as an important reminder of the personal pleasure to be found in plant collecting and the role of the garden as a multisensory space, even when it also had a scientific function.

This sensory approach does not mean that Fothergill's religious relationship to nature was eclipsed. His Quaker beliefs were reflected in his letters as he urged Bartram, "in studying nature forget not its author."[59] Plant collecting could then have a very personal meaning created through an intertwining of religious feeling, desires for pleasure, and serious scientific interest. It is clear that although Fothergill was an active collector of various productions of natural history, he viewed these as an investment for later life when he expected to have far more time than his snatched few hours with a lantern to enjoy them. In 1770 he wrote to Humphry Marshall, explaining, "Perhaps thou will be surprised when I tell thee one of my principal inducements to make . . . collections. It is that when I grow old and am unfit for the duties of a more active life, I may have some amusement in store to fill up those hours when bodily infirmity may require some external consolations."[60] This was, therefore, a collection designed for planned future leisurely enjoyment as much as to satisfy current scientific curiosity.

The Garden as a Botanical Clearinghouse

As we have seen, Fothergill's garden provided a place for growing plants collected from beyond British shores. Often these were the products of expeditions that were co-funded with fellow physician and botany enthusiast, Pitcairn. As noted above, royal botanic gardens such as Kew, under Banks, were already starting to provide a central "clearinghouse" where specimens as well as ideas and knowledge could be located and exchanged, but this term could also be applied to many other gardens, including those owned and developed by medical physicians.[61]

Like Fothergill, Pitcairn (fig. 2.5) bought a rural estate in Islington, which was then a village just outside London. There he developed a botanic garden that was well enough known for Mrs. Delany to visit it on her 1779 botanic excursions in London and to describe it as "Dr Pitcairn's botanical garden."[62] One of the earliest accounts is recorded by John Nelson in 1811 in his *History and Antiquities of the Parish of Islington,* in which he wrote, "About 30 years ago, Dr. Wm. Pitcairn began a botanical garden, behind the house in which he resided (now Mr. Wilson's), opposite Cross-street, and which he cultivated till his decease: this continues to be one of the finest

Fig. 2.5. Mezzotint of William Pitcairn as president of the College of Physicians, by J. Jones after Sir Joseph Reynolds, 1777. Wellcome Collection, CC BY 4.0.

gardens in Islington, and is upwards of 4 acres in extent."[63] Sadly, little evidence remains to give an indication of its design, but there have been suggestions that it was likely to have been laid out on a plan modeled on Boerhaave's garden at Leiden.[64] Given that Pitcairn attended Boerhaave's lectures as a student, he would certainly have visited the garden as part of the botanical course, and it may well have influenced his own design and use of space.

This private garden in Islington was clearly of significance to those within Pitcairn's botanical network. In his lecture on botanic gardens for his students, Hope stated, "I must mention also others, Drs Pitcairn, & Dr. Fothergill who possess Gardens the next to this Royal Garden in goodness."[65] The royal garden Hope was referring to was of course that at Kew (plate 7), and he continued to note the central role this garden played within the web of plant collectors and spaces beyond the garden: "The attention of his Majesty is not alone in contributing largely for the Botanic Gardens but also in sending Missionaries to distant parts of the world for procuring herbs and seeds. His Majesty has also missionaries to gather all plants of a rare kind."[66] In a similar manner, Pitcairn and Fothergill also funded expeditions to gather plants that in turn were grown in a range of domestic, commercial, and institutional garden spaces, including Hope's own botanic garden in Edinburgh. Hope's reference to them may well have also represented a polite nod to key contributors to his own teaching collection.

Like Fothergill, Pitcairn was a major medical figure, but one who was accepted to the highest levels of the profession without the obstacles of being a religious nonconformist. He acted as the physician to the well-established Saint Bartholomew's Hospital, London, and was the president of the Royal College of Physicians, also based in London, from 1775 to 1785.[67] Like many of the other medical professionals discussed in this book, he was also a member of the Royal Society and was elected as a fellow in May 1770. His application established his botanical expertise, stating that he was a "Gentleman very well versed in all branches of Literature and Natural History, and especially distinguished by his application to Botany and success in rearing scarce and foreign plants."[68] Among those who proposed his application were William Hunter and Fothergill, cementing his status within a network of eminent physicians and scientists.

Like other physicians, Pitcairn also participated in professional and social networks. For example, both Cullen and Pitcairn were members of elite households at the start of their careers. Cullen began his career as the Ordinary Medical Attendant to James, 5th Duke of Hamilton, and Pitcairn was private tutor to the 6th Duke. In these capacities Cullen and Pitcairn met and forged a friendship with William Hunter—all Scottish medical men with "a love of books in common."[69] The signature of Hunter on Pitcairn's Royal Society Fellowship application signifies the importance of such networks for professional development.

The circulation of botanic specimens has always been particularly valuable for the creation and maintenance of botanic gardens. As we have already seen, Pitcairn was supplying plants to Curtis's new botanic garden. However, his garden also provided plant material for more established teaching collections. For example, Rembert notes that Pitcairn was able to help the Chelsea Physic Garden in the 1780s when they needed assistance, along with other establishment figures. The committee of the physic garden ordered that thanks were to be "given to Sir Joseph Banks, Dr. James Smith, Mr. James Dixon of the British Museum, and Dr. Pitcairn for their plants and seeds for the use of the garden."[70]

Pitcairn was also exchanging plants with John Hope at the Leith Walk botanic garden in Edinburgh. Some specimens were from his own garden, and others were obtained from local nurserymen in London. On December 22, 1777, Dr. Pitcairn wrote to Hope with a receipt of plants that he had sent, recording that "many of them are from Mr Lee for in private collections we have no numbers of trees & shrubs."[71] In the same letter he also stated that "I received your alpines which came safe & in good condition."[72] He then goes on to list the forty-one plants that he had sent to Edinburgh, and in return Hope compiled "a list of trees & shrubs wanted in the B. Garden and are to be got in the neighbourhood of London viz. from Dr Pitcairn, Messrs Lee and Malcolm."[73] Here Pitcairn's garden provided a valuable resource for the university garden along with the nurserymen's commercial gardens of Lee and Malcolm, between which Pitcairn acted as a broker for Hope. Again, there is no real distinction made by users themselves between the gardens, and they are all perceived as equally valuable in the circulation of material between the various spaces, although Pitcairn, perhaps in his more privileged role within the network, seems to be the one negotiating the transfer of plants from the London nurserymen to Edinburgh. As Easterby-Smith argues, London's nurseries "contained collections of new plants that rivaled those of private amateurs and public botanical institutions. Their exclusive contents made them significant as sites of new knowledge."[74]

As well as sending plants out from his botanic garden, Pitcairn was also actively involved in plant collecting. In this way his Islington garden became a botanic clearinghouse where plants were brought in from around the globe and then distributed to other places, including Edinburgh. For example, Banks described the active role of Fothergill: "In conjunction with the Earl of Tankerville, and Dr Pitcairn, and myself, he sent over a person to *Africa,* who is still employed upon the coast of that country, for the purpose of collecting plants and specimens."[75]

This reveals a network that extended beyond that of the medical profession to the landed gentry, and one that was forged by a shared interest in botany. It also raises questions of other activities related to plant collecting, such as the elision of indigenous knowledge in this colonial plant-collecting expedition. More focused research in this area could reveal the relationship between these expeditions and other colonial activities.

Fothergill and Pitcairn worked together as collectors as early as 1768 and were involved in funding a number of expeditions to the West Indies, the Alps, and Africa with various other wealthy botany enthusiasts. Thomas Blaikie, for example, was employed by them on the Alpine collecting mission and recorded in his diary that a package including 420 seeds was sent on November 1775: "specimints (sic) and seeds sent together in one box directed to Dr. Pitcairn Warwick Court Warwick Lane London."[76] They are also recorded together as donors of several plants to Kew gardens in William Aiton's *Hortus Kewensis,* which was compiled predominately by Banks's botanical assistants, Jonas Dryander and Solander.[77] The *Kewensis* first appeared in 1789 as a record of the newly expanding plant collection housed at Kew, and lists donors, like Pitcairn and Fothergill, but not those who actually collected the physical specimens and sent them back to British shores, nor any local knowledge acquired in the collecting process.

This may in part reflect that many of the plants and seeds that were collected would have been grown, acclimatized, and propagated before they were sent on elsewhere, often in private gardens. In 1778 Henry de Ponthieu sent a letter with a parcel of specimens from his plant-collecting expedition to the West Indies. He wrote that he had "sellected an assortment of Seeds for the Kings Garden—Dr Fothergill's & Dr Pitcairn's."[78] In this way these physicians and their private gardens can be seen to be facilitating the colonial botanic enterprise in a similar way to that of Kew and other more prominent botanic gardens.

This use of the garden as a botanic clearinghouse through which plants circulated can be seen vividly in this description from Lettsom of Fothergill's garden (plate 8):

> From America he received various species of Catalpas, Kalmias, Magnolias, Firs, Oaks, Maples, and other valuable productions, which became denizens of his domain, some of them capable of being applied to the most useful purposes of timber; and, in return, he transported green and bohea teas from his garden at Upton, to the southern part of that great continent, now rising into an independent empire: he endeavoured to improve the growth and quality of coffee in the West India islands; the Bamboo cane (*Arundo Bambos*) calculated for various domestic uses, he procured from China, and purposed to transplant it to our islands situated within the tropics.[79]

This undermines Fothergill's claim that he was really a sensual botanist and wanted a garden to enjoy on his retirement from practice, a desire that was repeated in Autumn 1772 to John Bartram, when he wrote, "I look forwards, and that it is not impossible but I may live long enough to think it proper to decline all business. Then an amusement of this kind will have its use to lessen the tediousness of old age, and call me out to a little exercise when subsiding vigor prompts to too much indulgence."[80] Despite this sentiment of looking forward to gardening at leisure, in reality he created an economic botany powerhouse, which could be financially valuable for the nation and the wider imperial project.

Bullfrogs and Tortoises

Along with the exotic plants, there was also a circulation of animals both from other countries and between British gardens. As noted earlier, there was the chameleon spotted by Mrs. Delany and her niece in the hothouse of Chelsea Physic Garden.[81] Similarly, in Fothergill's letters to Bartram he mentions exotic species such as bullfrogs and turtles. The bullfrogs (plate 9) are of particular interest as they suggest how animals might circulate via private gardens in the same way or even alongside plant specimens. In 1770 Fothergill thanks Bartram for two letters as well as "the box of plants, the cast of *Colocasia* [a type of yellow water lily], and the Bull Frogs alive."[82] He goes on to write that

> a place is not yet fixed upon for the Bull Frogs to be put in. In the meantime however they are kept in a shallow vessel of water, the bottom covered with moss, where they may either put their heads above or under the water as they like. We have now a severe frost, but when all this goes off they will be set at large somewhere and in safety. We have none of the kind in England. The King is acquainted with their arrival; also the *Colocasia,* and from who they come.[83]

A few months later, in March, Lettsom had to inform Bartram that the frogs were still alive and well but not yet delivered to the king.[84] He suggested this was due to the "present state of public affairs," presumably the growing tensions between Britain and America, and says he would find a place for them in his own garden. Nonetheless, in 1772 he wrote again saying that although he had sent a description of the frogs to the king, he had heard nothing in return. At this point he also describes their place within the garden and the issues of trying to keep such animals captive:

> In a little place where I keep a few gold fish I put the frogs and fenced it in, in such a manner as I thought they would be forthcoming whenever they were called for. A small communication, between the place I had allotted for them and a large canal, underground, and of which I was ignorant, afforded one of them the means of getting more liberty. The other is still a prisoner, is still alive, and my gardener who sees him frequently tells me he is increased in size.[85]

At this juncture he suggested that as he has heard nothing from the king he might let the other escape so that it might find the original escapee. However, in 1774, despite seeming to be sanguine about the runaways, he is writing again with a request:

> Please let him [William] know that I received the turtle in good health, and shall be much obliged if he will procure me a male and female Bullfrog. Mine are strayed away notwithstanding my best endeavours. If they are put in a little box of wet moss, they will come safe; at least I received a little American frog, the *Rana ocellata,* in a box of plants, filled with moss.[86]

Here, then, we have a variety of frogs as well as a turtle arriving at Upton alongside parcels of plants. However, unlike the bullfrogs, which were located in the garden, the turtle may well not have been so lucky. The minutes book of the Society of Physicians has a record of members being delighted that they were able to dine on turtle soup at meetings held in the Crown & Anchor on the Strand, although the origins of that particular turtle are unknown, as is the destination of Fothergill's own creature.[87]

Fothergill was not the only medical practitioner with exotic animals, particularly from America, in his garden. Lettsom, in his description of his vegetable garden, notes that "here are left to range among the vegetables several tortoises, which are become so familiar, as to attend regularly the gardeners at their meals, and eat the leaves they offer from their hands."[88] These are clearly an integral part of the garden's living collection. Lettsom explains that the age of one of these creatures was over sixty-three years and that it had originally been sent as a gift by Humphry Marshall of West Chester, North America (plate 10). As a child Marshall had marked the tortoise himself, so when choosing in later life to send this to Lettsom, it must have had particular personal significance. Again the living elements of the garden represent substantial ties formed within a network built on the gift exchange common within natural history collecting.[89]

Within his estate Lettsom also had more traditionally demarked areas, such as a small farm where he housed chickens and hens, as well as an aviary and a menagerie in which he kept rather more exotic animals, including squirrels, flying as well as ground squirrels, a bear, and a great white American owl.[90]

Fig. 2.6. John Hunter's house at Earl's Court, as imagined by the rival surgeon Jesse Foot in his own extra-illustrated copy of his 1794 work *The Life of John Hunter,* vol. 3, 1822. Note the fantastical two-headed beast in the background on the right, which highlights Jesse's depiction of its unreal quality and the use of the illustration to imply Hunter was acting in an immoral manner and playing at being God with his animal experimentation. Wellcome Collection, CC BY 4.0.

There are a few such tantalizing references, but with little documentary evidence, of other animals in the gardens owned by medical practitioners. The physician William Withering, for example, developed a botanic garden at Edgbaston Hall in Birmingham. Best known for his work on the foxglove (*Digitalis*) plant (plate 11) and its role in the treatment of heart conditions, it is perhaps unsurprising that he had a botanic collection.[91] However, there are also notes that he kept monkeys at the hall, while breeding cattle and dogs.[92] Similarly, the surgeon John Hunter had a great variety of animals in his garden at his country retreat at Earl's Court (fig. 2.6). Using Hunter's papers and other descriptions, Stephen Paget in 1897 compiled a list of animals that he believed were kept at Earl's Court:

> In a field facing his sitting room was a pond, where he kept for experiment his fishes, frogs, leeches, eels and river-mussels. . . . The trees dotted about the grounds served him for his studies of the heat of living plants, their movements and their power of repair. He kept fowls, ducks, geese, pigeons, rabbits, pigs, and made experiments on them; also opposums, hedgehogs and rare animals—a jackal, a zebra, an ostrich, buffaloes, even leopards; also dormice, bats, snakes and birds of prey.[93]

Although this was compiled a century after Hunter's death, and some animals may have been kept in locations other than Earl's Court or only referenced in Hunter's work, it gives a sense of the potential range of animals that could be found in gardens of this time. A greater discussion of the distinction between exotic and domestic animals can be found in chapter 5, but this underlines the extent to which these gardens housed a variety of animal species that arrived via similar networks, and sometimes literally alongside the plant specimens, in packages from around the globe.

Like many other aspects of gardens owned by medical practitioners in this period, this inclusion of a range of creatures may have reflected the extensive menageries and aviaries constructed by members of the landed classes.[94] Men such as Joshua Brookes made a lucrative living from supplying a whole range of creatures to anyone who could afford them from his Original Menagerie in London, which demonstrates the popularity of animal ownership during this period.[95] In 1791, Gilbert Pidcock, a traveling showman and later part owner of the Exeter Exchange menagerie, held an exhibition at the Lyceum in London, where over four hundred animals were exhibited in the Great Room, including a lion, a condor, a silver-headed eagle, an imperial vulture, a pelican, a rattlesnake, leopards, macaws, and a hyaena.[96]

Many of these birds and animals would have made their way to the aviaries and menageries created within the landscape garden, which were so common by the second half of the eighteenth century that they went unrecorded.[97] Among those who were keen animal collectors was Queen Charlotte, who had several collections housed at Richmond, Buckingham Gate, and Kew.[98] At Kew, William Chambers remodeled much of the landscape in the 1760s. Alongside his much more famous Chinese pagoda and various other garden buildings, he also included a Chinese-style aviary (plate 12), which contained "a numerous collection of birds, both foreign and domestic."[99] There was also a menagerie, in which we are told by Chambers were "kept great numbers of Chinese and Tartarian pheasants, besides many sorts of other large exotic birds."[100] This New Menagerie (plate 13), which has since been developed and is now known as Queen Charlotte's Cottage, also included a collection of kangaroos, with a population of nearly twenty by the time it was dispersed at the beginning of the nineteenth century, as well as cattle from Algeria and India.[101] Royal taste once again seems to have influenced professionals such as Lettsom, and there is clearly a strong interrelationship between fashionable spectacle and natural history interest in the garden of this period.

Although these may seem to be for pleasure rather than for any greater purpose, animals could also have been viewed as important scientific specimens. The work of John Hunter in comparative anatomy relied on his access to a range of specimens, although many of them were obtained as carcasses from showmen.[102] He also wrote a number of papers on animal subjects for the *Philosophical Transactions of the Royal Society*. One such paper, titled an "Account of an Extraordinary Pheasant," appeared in 1780 and described his investigation of a pheasant that had been presented to him by Pitcairn, who in turn had originally received it from Sir Thomas Harris—its extraordinary nature being a "hen pheasant with the feathers of a cock," which could not breed.[103] This demonstrates how animal specimens could move between members of a network in a similar manner to plant material, and even between the same people.

This interest in both living and dead specimens located within the garden space can be illustrated by the earlier example of the physician, naturalist, and collector Hans Sloane. Within Sloane's landscape he housed a "red-headed crane from Bengal, a blind Arctic fox from Greenland and a large greenish lizard from Malaga" as well as a beaver that probably came from the New World.[104] The beaver was of interest as both a live animal when splashing in the fountain of his London garden, as well as a dead body to be understood by dissection. On its untimely passing following a series of fits and finally being attacked by a dog, the beaver was dissected by Sloane's friend and neighbor Cromwell Mortimer. His account of the beaver in both its living and dead states was published in the Royal Society's *Philosophical Transactions* in 1733.[105] The close analysis afforded by watching the beaver in Sloane's garden over a period of three months allowed Mortimer to offer descriptions of "how her Food was Bread and Water; some Willow Boughs were given her of which she eat but little; but when she was loose in the Garden, she seem'd to like the Vines much having gnawn several of them as high as she could reach quite down to the Roots" and that "when she eats she always sate on her hind Legs, and held the Bread in her Paws like a Squirrel."[106] This scrutiny of the live animal was mirrored in the detailed account of the anatomical dissection. Similarly, the garden itself could provide the necessary space for detailed examinations of the natural world. For example, in 1720, William Stukeley and Dr. Douglas were recorded as dissecting an elephant, which had previously been shown as a spectacle in West Smithfield, on the lawn of Hans Sloane's London residence.[107] This use of the garden space for both living and dead animals is best exemplified by Hunter at Earl's Court, and this will be discussed in more detail in chapter 5.

Overall, the evidence suggests that, as with Fothergill's bullfrogs and Lettsom's tortoises, such animals in the early modern period were to be found as much in the ornamental garden areas as in specific menageries. So Mortimer describes how the beaver was "turned into a Fountain with some live Flounders," and was placed here "to bath three or four times a Week."[108] It is notable that animals, like plants, were viewed as both scientific specimens for study as well as living additions that enhanced the ornamental and decorative spaces.

Invisible Garden Hands

These gardens with their exotic residents were not just clearinghouses for plants, places of scientific inquiry, and delightful spaces for sensual botanists. They also offered opportunities for gardeners to be trained in new techniques essential for their role as custodians for these expensive, exotic, and newly arrived species. This expertise could then be used to command better-remunerated positions or travel to other parts of the empire. Christopher Smith, for example, worked as a gardener for Pitcairn in the early 1790s and via this route came to the attention of Banks.[109] In 1794 he was promoted to the nurseryman for Roxburgh at the Calcutta Botanic Garden in India, one of the British botanical outposts.[110] Roxburgh himself had trained as a surgeon in Edinburgh in the 1770s and would have therefore had connections with Hope and others within this influential botanic network.

In her study of commercial botanic networks during this period, Easterby-Smith has demonstrated how some gardeners, although by no means all, managed to move from lowly roles to intellectually and commercially valuable elite positions within botanic networks.[111] One reason for this, she posits, was the importance of hybrid expertise to the development of botanic knowledge in the eighteenth century. She argues that for botany

> "hybrid expertise" describes how knowing about the growth and living characteristics of a plant might contribute useful information to the botanical project of developing systems and classifications. Further, hybrid expertise was most likely to develop in situations where knowledge flowed in both directions: from scholars to gardeners, and from gardeners to scholars.[112]

The experimental nature of garden spaces then allowed some gardeners to rise through the ranks if they demonstrated this necessary hybrid expertise. In this way the vital space these gardens provided for the creation, development, and dissemination of both scientific and practical knowledge becomes apparent.[113]

There were also close interconnections between medical and gardening roles beyond the patronage of physicians of the role of gardeners in medical training. One key example of the blurring of these roles can be seen through the relationship of Archibald Menzies to Hope in Edinburgh, and to Fothergill and Pitcairn in London. Born in 1754 near Aberfeldy, Perthshire, into a family of gardeners, Menzies started his gardening career at

Castle Menzies, owned by Sir Robert Menzies, 3rd Baronet of Nova Scotia.[114] Half of the twenty-one gardeners who worked there were members of the Menzies clan, and four of Menzies's brothers remained gardeners for the whole of their working lives.[115] The familial network offered up further opportunities for Menzies as his brother William, already employed as a gardener with Hope, presumably arranged for him to work at the same Leith Walk garden in around 1770.[116]

As well as working in the garden, Menzies collected Scottish flora for Hope, and in 1778 he also collected plants for Pitcairn and Fothergill from the Highlands.[117] This early relationship with the eminent London physicians and their botanical gardens was likely to have been brokered by Hope, who was already exchanging plants with Pitcairn, as we have seen earlier. Hope was a key figure here and clearly encouraged those who worked for him and demonstrated promise or enthusiasm to develop their own botanical and medical expertise. Like Fyfe, Menzies also attended medical lectures alongside his work as a gardener and plant collector. These classes were likely to have been subsidized by Hope and reflect both the important role of the Edinburgh botanic garden in training expert gardeners and Hope's own role in encouraging and perhaps even funding talented men to take up medical practice.[118]

This early beginning established Menzies as an ideal candidate to become a ship's surgeon, as he had both the medical and botanical expertise that would be useful on voyages to new lands. As noted above, ships' surgeons were key actors in the collection and development of natural scientific knowledge.[119] Hope's hand can be seen in Menzies's appointment on the *Prince of Wales* expedition. Writing to Banks, Hope noted that Menzies was "early acquainted with the culture of plants and acquired the principles of Botany by attending my Lectures" before serving for several years as surgeon's mate on a naval vessel on the Halifax station, where he "paid unremitting attention to his favourite Study of Botany."[120] This combined knowledge of botany and medicine then made Menzies ideal as part of the ship's team for exploratory voyages, as he could both treat any medical issues on board as well as collect and catalog specimens when on land.

However, even without such specialist medical training, gardeners could be employed on these trips solely for their botanic understanding. For example, David Nelson, a gardener from Kew, was employed by Banks as a "civilian supernumerary" on Captain Cook's *Discovery* voyage.[121] His functions were limited, but he was still considered useful for the expedition, although this may reflect Banks's own belief in the superior nature of horticultural knowledge.[122] Again, botanic networks were important in facilitating appointments. In this case James Lee, the nurseryman (plate 14), informed Banks that Nelson was "a proper person for the purpose you told me of, he knows the general run of our collection of plants about London under-stands something of botany but doe's not pretend to much knowledge in it."[123]

As well as active roles in plant collecting and moving between gardens, ships, and the wild, gardeners played a vital role in maintaining the collections of busy physicians. Lettsom complained in 1795 of a life spent traveling around London in carriages between appointments. Writing to Dr. Watson, he bemoaned how "as I live in carriages, seldom having less than three pair of horses a day, and neglecting my meals, except once a week that I dine with my wife, I have some time to preserve my correspondence, having always, in the carriage, pen, ink, and paper, to amuse myself, if I do not amuse my correspondents."[124] In fact, the rare moments in which physicians could enjoy these private gardens or indulge their interests were highlighted by the biographer and dissenting minister Joseph Towers, in his *Life of Fothergill,* in which he argued that natural history "affords the greatest instruction and recreation with the least exercise of the mind: it is, therefore, well adapted to the pursuit of a medical man, whose moments of seclusion are rather snatched from time by watchful diligence, than enjoyed from actual leisure."[125]

Given the busy lives of medical practitioners, gardeners were often left in charge of the botanic collections at physicians' country estates, although in Lettsom's case perhaps also with the oversight of his wife. According to his letters she preferred living in the country at Grove Hill, so she would have been present on the estate.[126] In this case, as Briony McDonagh has discussed in detail in relation to the management by women of other landed estates, it may well be his wife who was managing Grove Hill, although given the lack of records this is currently pure speculation.[127]

Fothergill, however, never married and lived mainly with his sister at his London town house, spending the summer with her from 1765 onward at an even more rural estate, Lea Hall in Cheshire. In Fothergill's obituary in *The Gentleman's Magazine,* Lettsom recorded that the elder physician rarely saw his Upton estate, as "he could visit only on Saturdays during the winter, and but rarely in summer, and in which fifteen men were constantly employed."[128] Corner and Booth also note that over time Fothergill had increasingly less time to supervise the garden developments, and that he "was sometimes observed there in the dead of night, lantern in hand, viewing by its glimmer its botanical treasures."[129]

In these rural estates with their absent owners, gardeners then were crucial in managing them, with or without supervision from a family member or other members of the household. As Easterby-Smith explains, finding an expert gardener who could nurture these expensive and exotic specimens was essential. She highlights how in the 1760s Peter Collinson noted with frustration that after he had supplied people with American plants, the next letter he received asked, "Pray sir, how and in what manner must I sow them . . . my gardener is a very ignorant fellow."[130] Horace Walpole also appears to have had problems with one of his gardeners, whom he accused of reducing "my little Eden to be as nasty and barren as the Highlands."[131]

Similarly, Fothergill wrote to John Bartram in 1768 relating the problems he was having with a head gardener who was less than able. He thanked Bartram for "a box of very curious plants which I received some time ago, and which are most of them prosperous, and all of them would have been so had my gardener taken the care of them he ought for they came in a very prosperous condition."[132] Fothergill received so many plants from around the globe that he developed a procedure so that new plants were registered by gardeners on arrival with information of where they had come from, any name that came with them, and where they should be grown at Upton.[133] In this way the gardeners of domestic collections performed important roles as catalogers, technicians, and expert horticulturalists, just as those working in institutional botanic gardens outlined earlier.

This importance placed on gardeners was no doubt heightened by the fact that Fothergill was rarely able to visit his own garden. He wrote to Bartram that he "ought not to think of increasing my collection for my leisure to attend to it seems to lessen every day—for on one occasion or another so many people seem to have claims to my assistance that I have less leisure than ever."[134] Presumably he had either taken on more staff or changed his head gardener due to the loss of plants, including a Pittsburgh iris sent by Bartram, as he also wrote that he now had "an able young natural gardener to take care of it, and though I see it not once a week now, yet when I do see it, it is always with so much satisfaction that I cannot relinquish it but live in hopes of enjoying it one time or another."[135] Although difficult to discover any detail regarding the gardeners themselves, it would seem that Fothergill had an expert gardener and botanist in the 1770s, one Mr. John Morrison. His death was recorded in *The Gentleman's Magazine* of April 1781 as "an ingenious botanist and principal gardener to the late Dr Fothergill."[136] As already discussed with the account of Williamson at the Leith Walk garden, expert gardeners were valued for their skill, and their often-hidden labor was essential for the success of such gardens.[137]

There was also a continual tension between the desire to have a garden and the lack of leisure time in which to enjoy it, hence Fothergill's often expressed desire that he was creating a space that he could enjoy in later life. Sadly, Fothergill never got the opportunity to leisurely enjoy his country estate at Upton in the way he wished, as he worked ceaselessly until his death in 1780. While he was alive, his garden, with its exotic animals and plants, of which thirty-four hundred were recorded as coming from warmer climes, reflected the ways in which colonial activities and the labor of those at home and abroad enabled such domestic collections to thrive.[138]

Future research will hopefully enable a fuller picture of these endeavors to emerge and reveal the ways in which our garden collections today have longer histories entwined with larger global and local narratives. By looking beyond the owners and designers of eighteenth-century landscapes and the visual appeal of their gardens with their beautiful displays of flora and fauna, we can perhaps gain a sense of their reliance on exploitative economic trade routes and less privileged human labor. Where we see the word "exotic," we should also see the actions of a colonial power.

THREE

For "Curiosity and Instruction"

Visiting the Botanic Garden

A GARDEN VISITOR ARRIVING at Grove Hill in the 1800s would have been given access to Lettsom's rural estate via the local Camberwell postman, who opened the gates in return for free accommodation in the garden lodge.[1] This innovative arrangement was established by Lettsom because his busy medical practice kept him in the city and prevented him from greeting visitors himself.[2] Once inside the gates, visitors would have been able to wander paths that wound through the extensive shrubberies and orchards, admiring the botanical specimens and newly imported American shrubs and trees (plate 15). These rare and exciting new plant introductions were an attraction, and the desire to see them was aligned with the growing fashion for garden and country-house visiting in the eighteenth century.[3]

The craze for the "English" style of garden design meant that touring estates was not limited to a domestic visiting public. From the middle of the eighteenth century, many foreign visitors also included a tour of British gardens on their itinerary.[4] There is extensive research on this growing leisure activity, but the inclusion of more modest gardens created by professional men, such as Lettsom, on this tourist circuit have been overlooked in favor of those more famous elite landscape gardens, such as Stowe in Buckinghamshire and Stourhead in Wiltshire.[5]

Whereas Lettsom's estate was only around ten acres, the famous Stowe landscape garden, under Viscount Cobham, had grown to around one hundred acres by the early eighteenth century.[6] As one of the largest and finest of this highly fashionable style of landscape park, it was unsurprisingly one of the most popular gardens for visitors, and a guidebook was produced in the 1740s for keen visitors by a local bookseller, Benton Seeley.[7] Alongside guidebooks for individual places, gazetteers containing lists of gardens of interest were also produced, including the mid-eighteenth-century *A Short Account of the Principal Seats and Gardens in and about Richmond and Kew,* which featured descriptions of seventeen properties that could be visited together within the same area.[8]

These publications along with the circulation of prints delineating grand houses and gardens fed the curiosity of the public. Those who owned remarkable or newly improved gardens could increasingly expect to be visited. In 1732, William Harper, chaplain to Lord Cholmondeley, wrote, "Whenever we hear any remarkable Seat very much commended, the first Question generally ask'd is, What *Gardens* has it?"[9] From the early eighteenth century it is clear that gardens were considered a draw and that anyone who presented themselves as a respectable visitor expected to be given entry.[10] A telling example is the experience of the Honorable Mrs. Boscawen, a well-connected, literary hostess. In October 1776 she wrote to Mrs. Delany regarding her visit to Luton Hoo:

> My aim was to see that delightfull conservatory *in particular* and the garden *in general.* As I pass'd the *castle-gate* in my way to the town, we enquired of the porter about seeing the garden, which he said we might do, and come in there, only keeping the gravel road, which would lead directly to the garden: it did so, and there I entertain'd myself highly above an hour; the gardener more civil and agreeable than ever I saw one, the conservatory more delightfull.[11]

Here we have the traveler arriving and being allowed to see the gardens without making prior arrangements, with the porter employed at the gate fulfilling the same gatekeeping function as the rather lowlier postman at Camberwell. For the gardener, such opportunities to demonstrate knowledge and agreeability could be rewarded with a generous financial tip, which could be important in supplementing what was often a low income. The estate staff were crucial for managing the garden visiting experience and could use it for their own financial advantage.

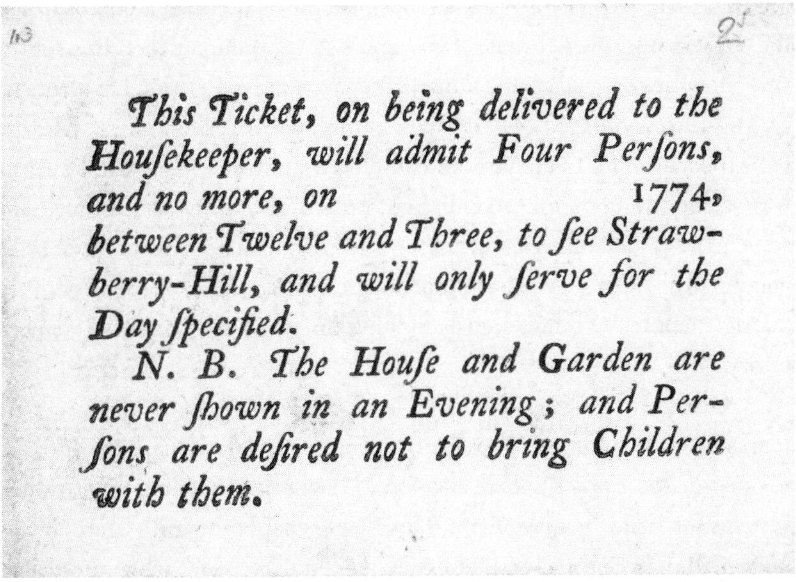

This Ticket, on being delivered to the Houſekeeper, will admit Four Perſons, and no more, on 1774, between Twelve and Three, to ſee Strawberry-Hill, and will only ſerve for the Day ſpecified.

N. B. The Houſe and Garden are never ſhown in an Evening; and Perſons are deſired not to bring Children with them.

Fig. 3.1. Entrance ticket including the rules for visitors viewing Walpole's house, Strawberry Hill, in 1774. Courtesy of The Lewis Walpole Library, Yale University, folio 336G.

This growth in tourism brought with it new problems. Daniel Defoe, the writer and journalist, on visiting the gardens at Wanstead in 1722 was told that "it has been the general diversion of the citizens to go out to see them, till the crowds grew too great, and his lordship was obliged to restrain his servants from shewing them except on one or two days in a week only."[12] Similarly, Horace Walpole, Whig politician and garden designer, wrote in 1783 about his Gothic house, Strawberry Hill, based like Grove Hill on the outskirts of London at Twickenham, that "I am tormented all day and every day by people that come to see my house."[13] In order to manage visitor numbers, Walpole developed his own ticketing system with rules for admission as a way of managing the public interest in what was a private residence (fig. 3.1).[14]

Ticketing, however, could be unpopular with visitors. On the same tour of gardens as the successful trip to Luton Hoo in October 1776, Mrs. Boscawen had a very different experience at Lady Diana Beauclerk's house on Muswell Hill, describing how "tho' we met her ladyship taking an airing, and that Mr. Beauclerc was in town, yet they would not admit us to see the conservatory (which was all we aspir'd to) without a ticket. Resistance you know, always makes one more obstinate, so Mrs Leveson has wrote to Sr Joshua Reynolds to beg *he will obtain* this necessary passport."[15] Clearly, the expectation that elite visitors could arrive unannounced and be let in was starting to come into conflict with the desire of owners to maintain some privacy. The blurring of public and private space was no doubt becoming problematic as visitor numbers grew and the use of tickets or other management techniques, such as limiting the hours for visitors, were necessary.

Given the broad range of fantastic and novel plants to be seen at places such as Upton House, owned by Fothergill, it is perhaps not surprising that medical practitioners also found themselves needing to introduce systems for visitor management. Banks states that Fothergill's "garden was known all over Europe, and foreigners of all ranks asked, when they came hither, permission to see it; of which Dr Solander and myself are sufficient witnesses, from the many applications that have been made through us for that permission."[16] This suggests that Banks and Solander were acting as intermediaries for those wishing to gain admission to Fothergill's gardens, and that visitors were traveling long distances to view the estates owned by the professional classes with an interest in botany, as well as the more elite landscapes, such as Stowe. Gardens like Fothergill's, Pitcairn's, and Lettsom's may also have been easier to access than those of the landed gentry due to their geographic location on the edge of cities, which may also have increased visitor numbers. However, it is clear that the fashion for seeing exotic botanical introductions was as important for visitors as experiencing newly designed landscapes.

We get a sense of some of the issues that came with owning a garden filled with exciting specimens by considering this narrative of a visit to Fothergill's garden. Joseph Cockfield in 1771 wrote that "Dr Lettsom is to visit me to-morrow to look at the Tea plant [plate 16], which is in our friend Fothergill's curious collection; we are to have a passport signed by Dr Fothergill, or most probably refused admittance. His plants are now become so numerous, that to prevent intruders he is obliged to have recourse to this method."[17] Here, visitors are described as "intruders," although in this case they are not described as interfering with the owner's peace as in Walpole's description—perhaps because Fothergill himself was mainly away in London working or in his country retreat in Cheshire. The term "intruders" may also point to concerns over the valuable nature of rare plants and their potential disappearance from the garden in coat pockets, if visitors were not properly supervised. The desire of visitors to take home elements of a collection was not just a problem faced by botanic garden owners; theft was also an issue for museum superintendents. The Ashmolean Museum, for example, included rules stating that only one group of visitors at a time should be allowed into the building in order to minimize the risk of items going missing or being moved.[18]

Although increasing the likelihood of the rare and the precious being removed, a physical, sensory engagement with such objects was still a central part of the eighteenth-century visiting experience.[19] There are clear parallels with the desire expressed by visitors to touch, smell, and sometimes taste museum objects in collections such as that of the Ashmolean, and with visitors wanting to physically experience living plants. As we have seen earlier at the Chelsea Physic Garden, John Floyer was tasting the plants and Mrs. Delany was handling them as well as removing parts of them for her work. Similarly, expert visitors, such as Celia Fiennes, an experienced traveler and writer, did not distinguish between the way they interacted with and observed museum objects and those investigations they made of botanic collections.[20] There was a shared approach to understanding collections which engaged the senses, and gardens can be seen within this wider context of the visitor experience, in which touch, smell, and taste could be just as important as seeing the collection.

Outside of this sensory experiential approach, which was tolerated despite the inherent risks, other problems were recorded due to the "wrong" type of visitor, often experienced by gardens that were located within easy reach of growing urban centers. Samuel Hellier, a member of the gentry, developed a garden at The Wodehouse near Wombourne in Staffordshire, which was very popular, with attractions including temples, follies, and a mechanical hermit, and also having an accessible situation on the outskirts of the growing urban center of Wolverhampton.[21] In July 1767, Hellier wrote to his steward John Rogers that "I hope the Wombourne people don't disturb my things and do mischief in the wood as they used to do. The gardner must watch it close. Particularly on a Sunday Nothing but Tag Rag and Rable come and play round the Upper Pool walls."[22] The role of the gardener as a caretaker and keeper is emphasized here; however, despite Hellier's letter, concern over the rabble amusing themselves in his woods continued, and by March 1769 the large numbers of the public visiting the estate meant that Hellier briefly restricted their access to his woodland garden. However, despite restricting the general public, he was, of course, not averse to letting Lord Stamford, owner of nearby Enville, and other members of the gentry visit.[23]

Similarly, although he professes concern about his gardener, Daniel, profiting from showing visitors around the garden, his decision to restrict numbers was based more on social class. In June 1772 he instructed his steward that "yt all strangers unless such who come in coaches and appear as people of Fashion shall be absolutely refused admittance. . . . I am excessively angry at Daniel for his takeing money or dareing to shew it anybody without permission from me. So pray have this order punctually executed or Daniel shall loose his place."[24] The gardener, it seems, had been showing around groups of people for economic gain which were not acceptable to Hellier, who remained clear that no people from nearby urban Wolverhampton, whoever they were, were to be admitted.[25] Visitors therefore were divided into those who were polite and fashionable and those who were seen as less desirable, often from the urban classes. It also again underlines the precarious nature of the employment of gardeners and the difficulties of raising income to supplement what were often meager wages.

This class-based discrimination, as well as the attendant problems of allowing visitors access, can also be seen in Philip Southcote's closure of Woburn Farm near Weybridge to the public after "savages, who came as connoisseurs, scribbled a thousand brutalities in the buildings."[26] In Dianne Barre's detailed description of Wombourne and its place on the visiting circuit for garden enthusiasts, she details the constant repairs and cleaning required to keep the estate looking its best and managing the effects of "dirty hands."[27] This constant battle is surely familiar to all who have managed a heritage site.

Other garden owners developed different methods for controlling the stream of visitors to gardens and separating the worthy from those considered less polite. In 1782, Hope published an advertisement outlining his system of managing visitor numbers to the Leith Walk botanic garden. He began by outlining that "much inconvenience has arisen from the crowds of promiscuous Company walking in the Botanic Garden by which the necessary work has been interrupted and proper distinction of visitors could not be made."[28] The promiscuous appellation alerts the reader to the fears Hope had regarding indiscriminate and casual visitors who would interrupt the main activities of the botanic garden—the teaching of botany to medical and other interested students—as well as not being able to tell the serious from the leisurely visitor.

As Hope's advertisement makes clear, his preferred public audience would be assembled from a mixture of the local gentry and serious botanists:

> On these and other accounts it has become necessary to admit none without an order from the Professor of Botany. By this regulation it is not meant to render access to the Garden difficult. Strangers, the Gentlemen of this county, the citizens of Edinburgh, and any person of knowledge or curiosity upon sending their names . . . will receive an order for seeing the Garden, between the hours of twelve and three and, during summer, at 6 in the evening every day, Sunday excepted.[29]

The need for a ticketing system as well as set hours to manage visitors indicates that botanic and small private gardens with interesting flora, like larger estates of the period, had become fashionable places to visit. Fothergill and Hope wrote that they were overrun by "promiscuous company" and "intruders," while at the same time welcoming more learned and gentlemanly guests.

Complex rules governed much country-house visiting, and different spaces had varying levels of access. Country-house libraries, for example, were often not part of the general visiting experience, as Lord Torrington discovered when he visited Belvoir Castle in 1789 and found the library locked on his tour—something he complained bitterly about, as well as a number of mistakes he felt had been made by the guide.[30] Although others, such as Croome Park, did allow visitors. The 1824 guidebook to Croome noted that the library was "open to inspection."[31] Other specialist readers were more welcome, though, particularly if they were guests of the family staying at the house. For example, John Skinner, a clergyman from Somerset, staying at Stourhead in Wiltshire, recorded how "he spent a couple of hours in the library before breakfast."[32]

Distinctions between types of visitor and the various spaces were not just restricted to private and institutional places. At Vauxhall pleasure gardens (fig. 3.2), a place open to all who could afford the entrance fee, elite accounts demonstrate that they had a very different and more privileged experience than the general public did.[33] Similarly, the British Museum, although outwardly a public institution, also made distinctions between different types of visitors, from the use of the rarefied reading room and use of specimens by the learned elites to the more public but still managed garden space.[34] Places such as these could be used by a range of distinct groups moving through them at different times and using them in particular ways, which could both confirm and consolidate divisions.[35] Like these seemingly egalitarian spaces of the public museum or pleasure garden, private or institutional gardens that appeared to allow access to a general visiting public did not always offer a comparable experience to those who were invited as guests of the owner, and some even deterred members of the lower classes from visiting. Whether by the porter at the lodge, a rejection of an application for a ticket, or some other means, the "wrong" kind of visitor could be discouraged.

Some of the issues caused by a broader visiting public originated with the proximity of suburban gardens to an urban populace. This was, of course, essential for medical practitioners who needed easy access to their town or city practice, or in the case of institutional and subscription gardens, a location close to students and subscribers. However, it is clear that

Fig. 3.2. Vauxhall Gardens was a space for the leisure and pleasure of a variety of social groups, as depicted and etched here by Thomas Rowlandson, from the *Microcosm of London,* pl. 88, October 2, 1809. The Elisha Whittelsey Collection, The Elisha Whittelsey Fund, 1959, Metropolitan Museum of Art.

such proximity created its own problems. Lettsom, always struggling to find time to visit his estate at Grove Hill, bemoaned that his practice kept him away from his plant experiments and added "that my villa being in the vicinity of London, numerous visitors ambulate my premises, and beg, or pluck up, some of the objects of experiment."[36] Here he explicitly links his garden's popularity to its location, and as Fothergill did, notes the problem of light-fingered visitors. In this case, though, it is not just about the monetary value of the specimens but the scientific loss of his experiments. This demonstrates the inherent friction between wanting to create a didactic landscape to encourage his particular style of gardening and the desire to have secure spaces for plant trials.

This growth in garden visiting, like that of museums and libraries, was also related to a new urban sensibility and a transformation of the urban experience through the opening up of new spaces for education and leisure.[37] The movement of urban populations to visit gardens within easy distance of their homes can be understood within the context of the wider development of civic space in the form of public walks, pleasure gardens, and, as I will discuss, semipublic botanic gardens. Although these more public horticultural spaces have been discussed in relation to urban polite culture, it is clear that domestic and university gardens also formed part of this civic development alongside museums, art galleries, and libraries.[38]

"The taste for natural History. is now become universal": Publicizing the Garden to Visitors

So how did audiences know about these private gardens with their exciting flora? In his own guide to Grove Hill, Lettsom states that he had received requests for information regarding the design of his garden after it had been featured in James Edwards's 1789 *A Companion from London to Brighthelmston, in Sussex.*[39] The *Companion* included topographical maps, plans, and views of country houses to be found on the road from London to what is now known as Brighton. It also acted as a knowledgeable guide to the natural history and antiquities of all the places to be seen from the road.[40] The *Companion* aided the tourist by including lists of inns, with timetables and details of transport, such as post chaises and stagecoaches.[41] It was, therefore, catering to a middling and upper-class audience interested in travel for leisure. That Lettsom's garden was listed alongside more elite seats of the landed classes suggests that Grove Hill, despite its modest scale, was perceived as fashionable and of interest to this newly mobile public. Other publicity for his garden was disseminated via articles published in magazines such as *The European Magazine, and London Review*. We can also assume that word of mouth played an important role too as people reported back on their visiting experiences.

It is also likely that Lettsom himself keenly promoted his own garden. He was one of the subscribers listed as contributing to the costs of publishing Edwards's *Companion*. This implies either that he wanted his garden to be advertised in order to accentuate his gentlemanly status, or that the inclusion of Grove Hill was a nod from Edwards to a patron. At the start of his own guide Lettsom is clear in his intention that his garden should be considered a model for those with smaller grounds in how to lay them out in a style that was "equally ornamental and productive."[42] This design, which combined those key eighteenth-century ideals of *dulci* and *utile*, will be discussed later.

The growth in novel exotic botanic specimens being grown in British gardens also led to a level of excitement that made them newsworthy as well as attractive to garden visitors. For example, the flowering of a *Rheum palmatum* (fig. 3.3) in a Norwich schoolmaster's garden was reported widely in newspapers in 1766, with the statement that it was the first time that the plant had flowered in Britain outside of the Edinburgh botanic garden, where it had flowered the year before.[43] A news report in *The Caledonian Mercury* of the "Great American Aloe" flowering at the Edinburgh botanic garden a year later was accompanied by set opening hours for visitors.[44] The fact that such events were worthy of reporting and also came with set visiting hours underscores the level of interest among the public in hearing about new botanic specimens. Such stories no doubt also encouraged a growing and increasingly mobile Enlightenment public to see the plants for themselves. From this perspective the type of garden in which the specimens were grown was of far less interest than the novelty of the species it

Fig. 3.3. Bottom right depicts the therapeutically and economically important Turkey rhubarb plant, *Rheum palmatum,* along with other plants of botanical interest, including a galled tree (left) and a Jamaica pepper tree (top). Colored engraving by Thomas Kelly, ca. 1827. Wellcome Collection, CC BY 4.0.

contained. As we have already seen, for the public looking for "curiosity and instruction," nurseries, private gardens, and institutional botanic gardens were all of equal interest as they allowed audiences the opportunity to view the flora for themselves.

Hope's Leith Walk garden also reflected a convergence between an increased desire for both polite and popular knowledge, with Hope attempting to manage this burgeoning interest by implementing his ticketing system. As one of his students recorded, Hope explained that "the taste for natural History is now become universal, & particularly for the study of

Botany. The no. of Interesting plants in this Garden is daily increasing. Coffee, Scammony, Camphor, the Indian pink, & the most wonderful of all, the moving plant are in great perfection."[45] Here the growing taste in natural history was linked to the greater range of plants increasing on a daily basis in the garden. As public participation in botanical investigation grew, botany itself became ever more intertwined with the cultural and social spheres of life, both fueling each other.[46] It wasn't just the introduction of novel and exotic species, however, that piqued such interest. The concept of "improvement," the economic trade in plants, and agricultural developments were also fashionable interests.

"For the use of the physician, the Apothecary, the student in Physic, the scientific Farmer, the Botanist . . . , the lover of Flowers and the Public in general": The London Botanic Garden

The need for accessible gardens to fulfill this burgeoning interest in botany and agriculture led to the creation of new botanic gardens that were accessible to the paying public. In 1810, Thomas Faulkner, in his topographical account of Chelsea and its environs, described the London Botanic Garden originally established by William Curtis (fig. 3.4) at Lambeth Marsh in 1777 as the first of this new kind of institution.[47] Curtis, an apothecary by trade and a Quaker from Alton in Hampshire, developed a successful practice in London (his share of which he later sold to his partner, William Wavell, in the 1770s).[48] This transition was vividly described by Dr. James Edward Smith, founder of the Linnean Society: "The street-walking duties of a city practitioner but ill accorded with the wild excursions of a naturalist: the apothecary was soon swallowed up by the botanist, and the shop exchanged for a garden!"[49] As with many of the medical practitioners mentioned in this book, botany was a passion for Curtis. As well as establishing the London Botanic Garden, he also founded *The Botanical Magazine; or, Flower-Garden Displayed* (begun in 1787 and the United Kingdom's longest continually running magazine). Here the paper version of the botanic garden complemented that of the physical subscription garden he created—a relationship that will be explored further in the following chapter.

Fig. 3.4. Portrait of the apothecary and botanist William Curtis, with copies of his *Botanical Magazine.* Yale Center for British Art, given by Lowell Libson in honor of Mrs. Rachel Lambert Mellon and the Oak Spring Garden Library.

Perhaps unsurprisingly, given their shared Quaker beliefs, medical network, and keen interest in botany, Lettsom and Fothergill played prominent roles in the establishment of the London Botanic Garden. In 1770 Curtis met Lettsom and Fothergill for dinner to discuss the concept of establishing a subscription botanic garden, in which all the species of plants that had been identified in the London area would be grown, with a central purpose to teach botany and the uses of plants, as well as providing a place for horticultural experimentation.[50] By 1819, Rees's *The Cyclopædia* was stating that as soon as the garden was established, Curtis's "pupils frequented his garden, studied in his library, and followed him into the wilds in his herborizing excursions"—all activities undertaken by students studying botany within institutional and professional settings.[51]

Unlike Oxford, Cambridge, Edinburgh, and Glasgow, there was no university with an associated botanic garden situated in London, so there was a gap for an enterprise that would provide a space for similar activities as those operating within more academic botanic collections. The only equivalent garden at this time based in the capital was the Chelsea Physic Garden, where Curtis had previously worked as a demonstrator.[52] The City of London's Worshipful Society of Apothecaries had established the garden in 1673, and it was predominately used as an educational facility for the training of apothecaries—people who prepared and sold drugs, which were often plant-based. The 1722 Deed of Covenant drawn up by Hans Sloane, the garden's benefactor, defined the garden as being "a teaching establishment for the appreciation of the 'power and glory of God in the works of the creation and that their apprentices and others may the better distinguish good and usefull plants from those that bear resemblance to them and yet are hurtfull.'"[53] Although, as noted earlier, other well-connected people with an interest in botany, such as Mrs. Delany, were able to visit the garden by the late eighteenth century, so by then it was presumably no longer solely for the use of the apothecaries.

However, the Quaker medical practitioners who were all keen botanists clearly felt that London needed a more publicly accessible botanic garden designed for a greater range of users. In his *Proposals,* Curtis argued that the London Botanic Garden (plate 17) was "designed for the use of the physician, the Apothecary, the student in Physic, the scientific Farmer, the Botanist (particularly the English Botanist), the lover of Flowers and the Public in general."[54] This is a much broader remit and is suggestive of a need for a wide base of subscribers to the garden in order to make it financially viable, but also an awareness of the growing popularity of botany and the acquisition of botanical knowledge by the public at large. The quotation also makes clear that from the very beginning there was an inclusion specifically of agricultural plants, which differentiates it from the earlier physic gardens with their initially more defined medicinal focus. Curtis himself produced a book describing English grasses and their agricultural uses, and within his *Proposals* he claimed that botanical knowledge "may be applied with as much advantage to agriculture as to any other science."[55] He also stressed his hope that the garden would "become productive of national utility."[56] This then places the subscription collection at the forefront of a widening of botanical and agricultural activities as well as audiences beyond the student of physic.

In addition to making the case for an accessible collection, Curtis also argued that the space would allow him to develop his own observational and experimental activities. In the preface to his *Flora Londinensis,* which cataloged and described the plants to be found in and around London, he stated that

> in order that he may obtain a more perfect knowledge of each plant; that he may see it in every stage of its growth, from the germination to the maturity of its seed; that he may compare and contrast the several species together; that he may make experiments to elucidate the nature of such as are obscure, or bring into more general use those which bid fair to be of advantage to the public; he is now cultivating each of them in a garden near the city, into which, by the kind assistance of his friends, he has already introduced, in the course of one year, about five hundred different species, including sixty of that most valuable tribe of plants the grasses.[57]

Here his stated aim was that the garden would provide both a public and a private resource based upon material supplied by his network of patrons and other botanical connections, again with an emphasis on the agricultural and commercial benefits of such an enterprise. It also emphasizes the importance placed on eyewitness accounts and the need to be able to observe a range of plants growing together in order to develop botanical expertise. The garden was again presented as a site for training the eye as well as the intellect.

The notes for subscribers stated that for two guineas a year they could access the garden on four set days of the week and each bring one guest with them. As a garden that relied upon subscriptions for its survival, Curtis was evidently hoping to fulfill a perceived need for an accessible botanic collection in London. According to Elliott, "Sixteen percent of the subscribers listed in 1790 were women acting in their own name, but many more would have gained admission as family members of friends."[58] This suggests a portion of the audience were women interested in developing polite knowledge, and highlights the fashionable nature of garden visiting, which encompassed seeing a whole range of landscapes, from elite private gardens through more raucous pleasure gardens to the scholarly and curious botanic collection. As Madeleine Pelling has made clear, and as the excerpts used in this book from Mrs. Delany's letters suggest, "in an age of botanical endeavour and plant collecting, classification and recording, the garden offered an environment in which women could learn and exhibit that learning."[59] This is more explicit in her examination of the Duchess of Portland and her female circle at Bulstrode Park and within Laird's focused account of the elite garden as a space for female scientific exploration and understanding.[60] This female interest in more domestic spaces would have encouraged traveling and exploring other collections, such as the London Botanic Garden, which was open to them via subscriptions in a way that perhaps other institutional botanic collections were not.

The subscription model for funding a new garden was not only promoted by London-based apothecaries and physicians. For example, in June 1779, *The Norfolk Chronicle* reported that "a number of gentlemen . . . are desirous of establishing a Botanical Garden near this City, upon the most liberal and perfect Plan."[61] The group argued that they could raise enough by "admitting Subscribers of One Guinea per year, to visit the Garden, from Ten O'clock till One, in the Winter Season, and from Two till Seven in the Summer."[62] As far as I can tell, nothing came of this plan, but it again highlights the fashionable nature of botanic gardens, the perceived scientific importance of such spaces, and a sense of regional and civic pride that such establishments could engender.[63] Other more successful subscription schemes that followed in the nineteenth century include the Liverpool Botanic Garden, opened in 1802; Hull Botanic Garden in 1812; and the Manchester Botanic Garden, developed on land purchased by the Manchester Botanical and Horticultural Society in 1831.

The function of these gardens followed those of the preceding century but with an attempt to make the spaces more accessible and open to a wider populace in the hope of increasing economic and cultural capital. Like the domestic and institutional gardens of the late eighteenth century, these were seen by the commercial classes as a way to encourage the dissemination of a utilitarian understanding of plants and highlight their potential economic importance.[64] These gardens can be seen as an extension to the polite culture of garden visiting developed in the former century as well as part of a newly forged civic and urban identity. Like the interlinking of museum, library, and garden collections seen in the earlier period, these new urban subscription botanic gardens should be viewed in conjunction with the founding of other elements, such as subscription libraries, zoological gardens, and literary and philosophical societies.[65]

This combination of the subscription botanic garden as a repository for knowledge, as well as offering opportunities for this new polite urban entertainment, can be seen in a letter published in *The Gentleman's Magazine* in 1810 by William Salisbury. Writing to highlight the successful move of the London Botanic Garden to Brompton and to encourage new subscribers, Salisbury stressed the role of the Dublin and Liverpool botanic gardens both as places for the dissemination of botanic knowledge and as spaces for rational recreation: "In order to combine rational amusement with study, they have Concerts of instrumental musick in the Garden on different evenings during the summer."[66] The subscription botanic garden, then, was perhaps a precursor to the public park, which emerged in the nineteenth century as an urban space designed particularly for rational forms of entertainment.[67] As early as 1810 Salisbury stated that he hoped the Brompton botanic garden "will in time be fully accomplished, by making it a scene of amusing and rational delight, as well as a repository of useful information."[68] It can then be argued that the gardens created by late-eighteenth-century medical practitioners and visited by an increasingly interested public paved the way for the development of urban parks and other gardens, via these semipublic examples funded by subscription, which were accessible to growing numbers of the public.

An extension of the concept of the subscription botanic garden as a place for education was outlined by the garden designer and prolific writer John Claudius Loudon in a paper titled "Hints for a National Garden," which he presented to the Linnean Society on December 17, 1811.[69] Loudon used this paper to call for the creation of a national garden, which could be used as a "living museum."[70] In many ways this was an extension of the types of botanic gardens already created in the eighteenth century, and Loudon's own career had some parallels with our cast of medical practitioners, including an education at the University of Edinburgh, where he attended classes offered by Professor Andrew Coventry in subjects including husbandry, horticulture, and "ornamental agriculture."[71] It is perhaps notable that from 1798 he was apprenticed to Messrs. Dicksons and Shade, nurserymen and landscape planners who were based on Leith Walk and thereby geographically close to the botanic garden established by Hope in the 1760s.[72]

However, as a man of the 1800s rather than the late eighteenth century, Loudon was more concerned with the generalist's use of plants for both utility and beauty and how the broader population might both contribute to and be educated by a public garden. In a shift from the gardens of our medical practitioners, this public garden would be laid out for the benefit of the public rather than designed for scientific botanists. He also argued that such gardens could be used to train the newly emerging professional landscape gardener in a manner that foreshadowed the gardens laid out by the London Horticultural Society (later the Royal Horticultural Society).[73] The society maintained a series of small "experimental" gardens from the early 1820s in Little Ealing, then Chiswick and various locations in Kensington, including one in the 1860s on land leased by the Royal Commission for the Great Exhibition of 1851, before developing the considerably much larger garden at Wisley, Surrey, in 1903.[74] This is still the RHS's central scientific and experimental horticultural base. The estate was donated to the society by the Quaker Thomas Hanbury, who himself had established a botanic garden at La Mortola in Ventimiglia on the Italian Riviera with his pharmacist brother, Daniel, in the 1860s[75]—thus demonstrating the longevity of links between medicine, education, science, and horticultural experimentation.

At the forefront of those calling for public gardens for horticultural education was Loudon, who argued to the other Linnean Society members that the new tastes in gardening could be communicated via different arrangements of garden design. For example:

> Specimens will be exhibited of different characters of surface, such as shaven lawn, smooth turf, forest or wild scenery, rough surfaces, broken Grounds, natural like roads, and walks, Groups, thickets; and above all picturesque pieces of water; that which judiciously introduced cannot be a greater source of improvement in the appearance of artificial Landscape. . . . In a word nothing would be omitted calculated to blend splendor, beauty and variety, with taste, science and utility.[76]

In many ways this combination of beauty with science and utility was simply a more public-facing version of the late Georgian gardens created by medical practitioners in this book. However, as a promoter of landscape gardening as a newly emerging professional field of work, we can also see Loudon here arguing for the use of an educational garden to train a new class of gardener.

Loudon remained faithful to the idea of public gardens as both educational and ornamental spaces, even when discussing cemeteries toward the end of his life. In 1843, Loudon wrote that "a general cemetery in the neighbourhood of a town, properly designed, laid out, ornamented with tombs, planted with trees, shrubs and herbaceous plants all named, and the whole properly kept, might become a school of instruction in architecture, sculpture, landscape-gardening, arboriculture, botany, and in those important parts of general gardening, neatness, order and high-keeping."[77] Here his "living museum" concept was simply transferred to a garden cemetery setting and had clear synergies with his designs for the Derby Arboretum (one of the earliest public parks in Britain, opened in 1840). The arboretum, donated by Joseph Strutt to the people of Derby, was designed by Loudon as a place that would encourage an interest in and educate the public in natural history and botany, which again places Loudon's work within a longer history of educational and scientific gardens, as well as parks as places of pleasure and leisure.[78]

Rather, it was the public-facing focus of his garden concept that distinguished it from the earlier botanic gardens, whether university, subscription, or private. In his treatise Loudon was at pains to make clear that the garden should appeal to more than the botanist who only needed specimens of plants for identification purposes. He wrote:

> Every one is pleased with a conservatory of large orange trees. How much will this gratification be increased by many species arrived at similar maturity; such as the Olive, Breadfruit, Cocoa, Palm, Plantain, Coffee, Pepper, Cotton, Indigo, Mahogany, Date, and a thousand others known only to general observers by name, or some useful property, and to botanists by sprigs in pots—what a source of gratification to curiosity, and of delight to the admirers of the bounty of nature![79]

Even though this sounds similar to our eighteenth-century garden with its collections of exotic material from around the world representing the fruits of empire, the exhortation to grow plants to their full size so that they can be appreciated more easily was clearly aimed at delighting a less scholarly audience. He goes on to suggest that systems other than the Linnaean classification would also help in the education of a less polite audience, arguing that "the Linnaean method, which though far preferable to all other systems for the purposes of the botanist, yet is compared with some arrangements far from being inviting to the general observer."[80] On that basis he suggested arrangements by types of plant on either natural or alphabetical systems.

There are other connections that link Loudon's idea to the earlier gardens explored in this book. Like our medical practitioners, he encouraged a sensory approach. In his design the outer circle would be comprised of the greenhouses and these "would be spacious to admit Carriages and when driving along its center in the Winter Seasons, surrounded by perpetual verdure, bloom, and fruit, jets of water, singing birds and a mild fragrant atmosphere, the effect would surpass any thing of the kind hitherto known, but in the regions of romance."[81] Here, then, he proposed a sensuous combination of scents, sounds, tastes and colors, which was intended to encourage a love of gardening and horticulture.

He also argued that it would have a strong agricultural application with a "specimen of each species and variety of livestock used in the husbandry of all the countries in our parallel of Latitude, with a repository of all the agricultural implements hitherto in use in every part of the known world."[82] This has clear parallels with the discussions of agriculture and gardens later on in this book in relation to improvement. The argument that this national garden would also need a library stocked with relevant books again suggests the interlinking of different collections for knowledge creation and dissemination.

So taking this long view, Loudon's work can perhaps be best understood as a bridge between the eighteenth-century botanic collection, which was partially accessible to the public but still aimed predominately at a politely educated elite with a deep knowledge of botanical science, and the public parks for rational leisure designed by the Victorians. These parks were also often developed in tandem with libraries and art galleries, alongside the more corporeal act of promenading, before the later emphasis on playgrounds and spaces for physical activities.[83] The late-eighteenth-century attempts at scientific and educational gardens that were accessible via ticket, subscription, or sometimes just by turning up at the gate were forerunners of not only today's specialist botanic gardens but also designed public spaces more broadly.

The influence of Loudon's ideas can be seen in the creation of hybrid botanic gardens, such as the Glasgow Botanic Garden, first proposed in 1817. These became common in the nineteenth century and were conceived as both public recreational spaces and scientific collections. In 1818 the *Companion to the Glasgow Botanic Garden* was published. This included a description of the aims of the garden, which can be seen as partway between the subscription botanic garden proposed by Curtis and the later nineteenth-century public park:

> To the agriculturalist, the horticulturalist and the medical man, the advantage of a Public Garden will readily be apparent; and

> even when viewed as a place simply of amusement and recreation, as a place where a natural and laudable curiosity may be gratified by the sight and knowledge of plants, of which we frequently hear and read, and the products of which we are in the daily habit of using, and as a place where a constant succession of agreeable or curious objects are presented to us, the Botanic Garden is well worthy of support.[84]

Following Loudon's example, the authors of the *Companion* argue that along with the usual Linnaean beds and more natural arrangements, "the different forest and other trees and shrubs, are grouped together, without any regard to classification, but so as to produce the most agreeable and ornamental effect."[85] The garden was also to be used by the medical faculty, and a lecture room and library for both students and the public were part of the original design. Here then was perhaps the embodiment of Loudon's "living museum" proposal, albeit without the animals and physical examples of different forms of landscape design.

FOUR

"Hints or Directions"

Reading the Doctor's Garden

VISITORS ARRIVING AT Grove Hill in Camberwell (plate 18), could navigate their way around the garden using both Lettsom's own guide, *Grove-Hill: An Horticultural Sketch* (first printed in 1794 and revised and reprinted under the slightly different title of *Grove-Hill: A Rural and Horticultural Sketch* in 1804), alongside the plant labels that marked individual specimens. As Pettigrew remarked, "Here every plant had its classical name distinctly given on its label; so that, with a manual of Botany in his hand, traversing these delightful walks, a person might with great facility have made himself a tolerable botanist."[1] Pettigrew clearly assumed that the garden visitor might bring his or her own botanic manual in order to develop botanical identification skills. This all suggests that for visitors the sensory experience of moving through the garden was likely to be supplemented with other written texts, such as guidebooks, manuals, and labels. This use of paper technologies has been widely discussed in relation to early modern science and botanic knowledge creation, but perhaps less so in relation to the "reading" of gardens themselves.[2]

This relationship between the physical materiality of the landscape and the written word is of particular interest when considering the use made by visitors of a variety of botanic collections. As Lettsom stated at the beginning of his guidebook to Grove Hill, he hoped that it would "assist the proprietors of country houses, in possession only of small allotments of garden ground, in laying them out."[3] This suggests that the main aim of the text was to aid visitors in their knowledge of the various plants and then potentially allow them to replicate Lettsom's style of gardening, marrying the ornamental with the productive, in their own domestic situation. He clearly felt that many houses with relatively small gardens could be improved with "respect to ornament as well as horticultural œconomy" and that his garden with its associated guide could act as a model.[4]

This didactic approach to encourage gardeners to make the most of their plots required a range of materials, from labels identifying new and exotic plants, many of which could potentially be economically productive, to guidebooks explaining the designer's approach and how it could be replicated. As Hunt has delineated, this desire for a garden to gain a response from its visitors helps substantiate the idea that "gardens fully exist only as a melding of an object (their physical ingredients) and a subject (the receiving, perceiving visitor)."[5] The garden then can only be fully understood by considering how owners, designers, and visitors tried to make sense of the materiality of the landscape. To ensure the "correct" response from visitors, there is often a need for mediation between the physical garden and the formation of meaning by the viewer. As we have seen, private elite landscapes such as Stowe were shown to visitors by those who owned or lived on the estate, or in their absence, knowledgeable workers such as gardeners, who could through their words interpret elements and direct attention to particular areas.[6] This could then be supplemented by written guides and other textual methods of interpretation.

As a garden is explored by moving through it, often on foot as "an interweaving of vision and movement," the textual works offer a method of telling stories and linking sometimes a singular plant to a broader narrative that extends beyond the specific time and place in which it is immediately encountered.[7] The various supplements then connect the particular to its wider context beyond the garden as well as revealing the specifics of the owner's interest. Such textual and verbal explanations were also a method of creating order out of chaos, or as Spary defines, a transition from "natural (brute) to social (member of polite society)," which was an essential conversion if owners wished to demonstrate their enlightened status.[8]

This is epitomized at Grove Hill, where the "social harmony" of Lettsom's dwelling with his Linnaean-arranged beds was contrasted with the chaotic and tragic narrative of George Barnwell. The murder by Barnwell of his uncle, as depicted in the popular eighteenth-century moral play *The London Merchant,* took place on the site that eventually became Grove Hill, before Lettsom purchased it. The narrative appears in both the guidebook and the poem devoted to Grove Hill by Thomas Maurice.[9] In the poem by Maurice this is used as a device to contrast the pollution of the "secret shade" and the murderer's "gleaming blade" with Lettsom as "humanity's and virtue's friend."[10] In this way the depiction of the approach to Grove Hill is both a physical description of the avenues of trees and a moral tale where Lettsom's house and garden represent an ordered, morally uplifting, rural world that has been transformed out of the disordered, chaotic character of the place.

This moralistic approach both reflected the eighteenth-century concept of "sentiment," which could be improved via access to cultural and scientific productions, as well as Lettsom's own Quaker preoccupations.[11] His moral and medical concerns are evident in his voluminous writings on a variety of subjects. As *The Gentleman's Magazine* memoir of Lettsom's life noted, his "writings are very numerous, as well moral as medical, and all of them discover the philanthropist and physician."[12] In 1801, for example, Lettsom published three volumes, which collated many of his works together, titled *Hints Designed to Promote Beneficence, Temperance and Medical Science.*[13] To give a sense of the scope of his writings, these included such tracts as "Hints Respecting the Immediate Effects of Poverty," "Hints Respecting the Society for Bettering the Condition, and increasing the Comforts of the Poor," "Hints Respecting Crime and Punishment," "Hints Respecting the Bite of a Mad Dog or Rabid Animal," and "Hints Respecting a Substitute for Wheat Bread." The garden in many ways can then be read as a physical manifestation of Lettsom's medical and philanthropic concerns, and his guidebook offers us an insight into the didactic nature of the landscape, as well as offering us a virtual tour of the ornamental and botanic offerings available to visitors.

Grove Hill: A Rural and Horticultural Sketch

Lettsom tells us that he produced a guide to his garden at Grove Hill due to the requests he had received for an account from "foreigners of taste and

Fig. 4.1. Visitors are depicted enjoying the landscape in this illustrated guide to the house and gardens of Blenheim Palace, Oxfordshire, England, first published in 1789. Apart from Sundays and public holidays, the site was open to visitors from 2:00 p.m. to 4:00 p.m. every day, according to the author. From William Fordyce Mavor, *A New Description of Blenheim . . . : A New and Improved Edition. Embellished with an Elegant Plan, etc.*, 8th ed. (Oxford: J. Munday, 1810). The British Library.

curiosity," which followed the inclusion of the description of his garden in Edward's travel guide. The publicity of houses and gardens of interest to tourists was part of a growth in printed material that by the end of the eighteenth century included travel journals, engravings, and other visual depictions of country estates available as prints, newspaper articles, and descriptions in periodicals.[14] However, guidebooks as a specific type were still relatively rare productions, and fewer than twenty elite country houses published their own guidebooks before 1815 (fig. 4.1).[15] This makes Lettsom's guide all the more intriguing, as his garden was not on the grand scale of a landscape such as Blenheim, Stowe, or Stourhead, yet he appears to have felt a desire to produce a detailed guidebook as early as 1794.

This may to some extent represent an attempt at self-publicity, although Lettsom's status as an elite physician was confirmed by this date and he was not in need of extra clients. It is also clear that he did not circulate the work for financial gain, as he printed the guidebooks himself and presented them to his friends, possibly as a way of cementing his status as an expert botanist and horticulturalist. This lack of a desire for personal publicity is exemplified in a letter sent to Dr. Walker in which Lettsom discussed the recent publication of the poems by John Scott and Maurice, which both described his garden. In relation to Maurice's ode, he felt that "it would please me, as one of the most elegant pieces of poetry I ever read, if it did not contain too many stories about me."[16] Similarly, he tended to publish much of his work anonymously, possibly in line with his Quaker principles. On writing to Sir Mordaunt Martin, a keen agriculturalist, in 1789 he stated, "I see no reason to affix any person's name, but entitle the paper, 'Hints or Directions for cultivating Mangel Wurzel with its Uses &c.' This would be doing good to the country, and is true patriotism."[17]

We could also read this suggestion of modesty in light of the fine line often trod by Georgian professional men such as Lettsom. Being perceived as self-seeking by your peers was frowned upon, and the aim was, instead, to be elevated within society through more modest means, such as a recognition of the development of a particular skill, expertise, or sensibility. The experience of early-eighteenth-century actor, botanist, writer, and apothecary John Hill would have been a salutary lesson to other learned gentleman in how their reputations could be lost through celebrity.[18] Often regarded as an expert and knowledgeable man in varied fields of endeavor, who published almost constantly through his life, he continually emphasized his own intellectual superiority. Hill was therefore shunned by his peers for a lack of acceptable conduct, manifesting itself in an absence of appropriate manners and modesty. His attempts at self-promotion as well as written attacks on others meant he was never easily accepted into the elite circles of the period. Lettsom himself was not immune to the censure of his peers. He fought a war of letters published in the press with the irregular practitioner Dr. Mysersbach, who promised to diagnose and treat patients from investigating their urine samples alone—a system Lettsom called into question for its lack of medical veracity.[19] This public spat was not always well received by other medical practitioners, and Lettsom may have learned that not all publicity was good publicity, particularly if you wished to be seen as a gentleman of modest sensibility.[20] Perhaps more importantly for medical practitioners such as Lettsom, there was also a risk of being perceived as a quack, or irregular practitioner, if they were seen to be partaking in self-promotion through means other than medical pamphlets and letters published in reputable outlets.

Rather, we might understand Lettsom's guidebook as less about self-publicity and more of an attempt to consolidate his status as a polite landowner and to share and promote his values as a Quaker physician within his network. He may, then, have been using the guide in a similar way to other guidebooks and descriptions of more elite landed estates. Written either by or for visitors, these texts helped to form new recognizable identities for country houses and transform them into places that could be read and understood more easily by the wider public.[21] The guide to Grove Hill certainly fits many tropes of eighteenth- and early-nineteenth-century literature depicting the fashionable landscapes of the day. Lettsom, for example, described the location as being placed upon a "picturesque hill," classical allusions featured in the form of a statue of Flora, and a patriotic note was struck with a memorial to Shakespeare.[22]

The reader was also directed to consider more mercantile themes, such as the enlivening prospect of commerce and productivity, with the description of the Thames with its "floating forest of ships" and the five "Telegraphs" that could be seen from within the bounds of the garden.[23] The picturesque nature of the landscape in this description was enlivened by illustrations and descriptions of this commercial and technological activity. This places Lettsom's garden into dialogue with the late-eighteenth-century landscapes created by wealthy merchants, such as those based in Bristol in the same period. The owners of them, many of whom were also Quakers, could in some cases even view their own ships as they arrived at the city from viewpoints in their newly improved gardens.[24] Lettsom pronounced that his view from Grove Hill combined "naval grandeur and rural elegance, no where equaled in the world, being indisputably the richest scenery that ever was afforded to the sight."[25] These combined interests in beauty and industry were also reflected in the tablets that were sculpted in *alto-relievo* on the exterior walls of the house. These represented a range of eighteenth-century sensibilities: arts, commerce, peace and plenty, woollen manufacture, sovereignty of the laws, and truth.[26] This, then, was a house and garden designed to be read as a productive and artistic enterprise that reflected the moral sensibility of its Quaker owner.

From his letters it would seem that the guidebook was initially conceived as a volume dedicated to a discussion of agricultural matters. In 1790, Lettsom wrote to Martin stating, "I mean to pursue another work, to be entitled 'Meditations and Reflections at Grove Hill' in which I shall make use of thy name, with acknowledgments, on various subjects in agriculture."[27] Given their long-running correspondence regarding agricultural developments, and in particular the growing of mangel-wurzle (a variety of beetroot/turnip), this is perhaps unsurprising. However, the perpetually busy Lettsom wrote again in 1791 repeating his claim that "I hope 'ere long to get some evenings, from other avocations, to complete a volume, to be entitled 'Reflections at Grove Hill,' in which Mangel Wurzel shall shew its countenance. It will be in company with potatoes, and some other subjects of vegetation: but, alas! I have been pursuing Time all my life, but never yet could get him by the forelock."[28] The overworked physician clearly struggled to find the time to write his work, and by the time he did, it still retained some Virgilian nods to agriculture, but was less focused on the vegetable garden and more concerned with demonstrating how agriculture and ornament could be combined.

The guidebook is not the only extensive written description we have of the garden. Maurice's long-form laudatory poem *Grove-Hill, A Descriptive Poem,* with illustrations extolling the virtues of Lettsom's garden, was published in 1799. This, like the guidebook, followed a form familiar to those interested in garden design in the period. Complimentary epistles to landscape gardens were a common descriptive form, and Maurice's ode can be seen in a long line of such celebratory verse, including Richard West's *Stowe, The Gardens of the Right Honourable Richard, Lord Viscount Cobham* (1732) and the anonymously penned laudatory poem *A Ride and Walk Through Stourhead* (1780). Maurice himself had already written a poem extolling the pastoral beauties of the famous Hagley landscape garden in 1776, which was preceded in the printed version by a description of the circuit to be taken by visitors.

Within the Grove Hill guidebook itself, Lettsom included a shorter poem by Scott dedicated to the gardens, which emphasized the picturesque scenery beyond Grove Hill, the skill of Lettsom as a physician, and the importance of the garden as a place of sociability. Poetry, unsurprisingly given its importance as a form by which garden design was discussed, promoted, and altered, was also used as a form to express the various ways in which Grove Hill intersected with the wider context of both its own geography and the sensibility of its owner.[29]

These texts also place the garden firmly within the wider context of empire, industry, and the economic utility of the plants being grown and experimented upon at Grove Hill. Maurice's poem mentions Arabian groves, Columbia's fields (presumably a reference to the plants from South Carolina), and India's clime, as well as stating, "Commerce makes the globe's vast wealth her own."[30] Again, as in earlier chapters, we need to view such gardens as products of empire and, in this case, one that was celebrated as a microcosm of British domination of large swathes of the globe. Economic wealth was bound up in botanic material at this time, and this conspicuous consumption by Lettsom, particularly as a Quaker medical practitioner, was only tempered by his argument that this was also a utilitarian space that was important for national wealth creation. This need to direct the reader or visitor to the ideal method of understanding the garden may have earlier precedents. For example, Louis XIV developed several different versions of his guide to the gardens at the Palace of Versailles, Paris, in an attempt to control both their reception and the itinerary taken by visitors.[31] Such texts, therefore, dictated how the landscape should be received and understood by the viewer.

In Maurice's opening "Preface" he stated that he had been struck by the scenery and beautiful landscapes at Grove Hill and that this experience had caused "an instantaneous desire excited in his mind to express the sentiments he felt in poetry."[32] Lettsom also recorded that Scott had a similar response after spending an afternoon with the physician at Grove Hill; an experience from which he "broke forth" inspired to write his "de-

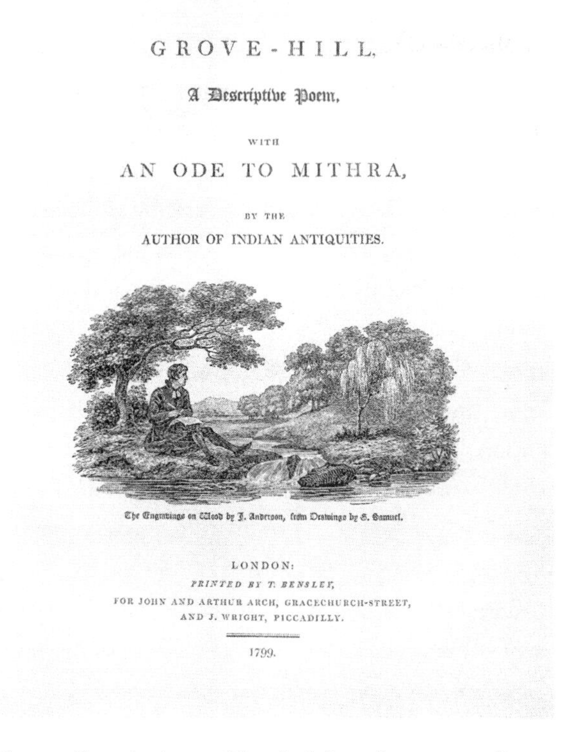
GROVE-HILL,

A Descriptive Poem,

WITH

AN ODE TO MITHRA,

BY THE

AUTHOR OF INDIAN ANTIQUITIES.

The Engravings on Wood by J. Anderson, from Drawings by S. Samuel.

LONDON:

PRINTED BY T. BENSLEY,

FOR JOHN AND ARTHUR ARCH, GRACECHURCH-STREET,

AND J. WRIGHT, PICCADILLY.

1799.

Fig. 4.2. Frontispiece to Maurice's long-form poem *Grove-Hill, A Descriptive Poem, with an Ode to Mithra* (London: Printed by T. Bentley, 1799). Wellcome Collection, CC BY 4.0.

scriptive eulogy."[33] Both poems are depicted as arising from instantaneous emotional responses caused by this landscape, which was consistent with the concept of picturesque sensibility of the time.[34]

The poem (fig. 4.2), however, also acted as a visitor's or virtual guide to the different areas of the house, garden, and surrounding landscape, with sections referring to specific features, including "The Lawn," "Shakespeare's Walk," "The Apiary," and "The Telegraph." Perhaps, as it was sponsored by Lettsom himself, this poem does more than just discuss the garden features.[35] Rather like the guidebook and the materiality of the garden, and to some extent Scott's short epistle, it also highlighted Lettsom's

Fig. 4.3. Engraving depicting the statue of Hygeia and the Fates, highlighting Lettsom's medical status. From Maurice's *Grove-Hill.* Wellcome Collection, photo by author.

status as a botanist, scientist, and philanthropist. In many ways this garden was designed to be read through the emblems of the Enlightenment in a similar manner to the earlier Stowe landscape.[36] At Grove Hill the educated viewer was encouraged to read the symbolic nature of the physical landscape features through a shared polite, cultural understanding, as well as a more directed approach reinforced by the guidebook.

The philanthropic nature of Lettsom's work was recorded in Maurice's poem through somewhat sycophantic lines, to modern ears at least, such as: "But chief the suffering poor thy worth proclaim / And call down blessings on thy honoured name."[37] Similarly, his work as a physician was represented in emblems physically within the garden and described in the texts, such as a statue of Hygeia (Greek goddess of cleanliness and hygiene and daughter of Asclepius, god of medicine) repelling the Fates (fig. 4.3), and a carved relief of the Great Pyramid of Egypt with the figure of Isis of Sais as a serpent. In explanation for the inclusion of the latter, Lettsom records that Isis "was supposed to be the revealer of the mysteries of Nature, and to have been an universal benefactress; but more especially to have presided over Medicine."[38] These were both symbolic of Lettsom's status as a physician and reminded the visitor and reader of his profession. Similarly, the relationship between medicine and the plant material of the garden itself was drawn by Maurice as he described how the air was scented by flowers so that "health comes wafted on each vernal gale."[39] The sensory nature of the garden as a healing space, particularly through floral scents, was highlighted here and encouraged the reader to understand the landscape as a whole as a medicinally grounded and therapeutic space, reflecting the healing abilities of its owner.[40]

The garden as described in the poem represents a physical manifestation of Lettsom the man. As well as the references to medical practice, Maurice praises Lettsom's own skill as a gardener, ignoring the crucial role of the team of gardeners based at Grove Hill and the physician's busy London practice, writing, "Rear'd by thy hand each plant more vigorous grows, / And lovelier far in yon rich garden blows."[41] There are also allusions to Lettsom's ability to subjugate the seasons, presumably through the use of the range of greenhouses, and his patriotic domination of nature using scientific methods. As well as his command over the climate, there are detailed descriptions of both the library and items in the museum (fig. 4.4), which are material examples of Lettsom's status as a learned man, and references to the "Telegraphs" (fig. 4.5) in the borrowed landscape beyond where "impatient Science leaps th' opposing mound," and the astronomical observatory, which gives Maurice an opportunity to praise Newton and thereby elevate Lettsom's own scientific standing.[42]

Aside from this depiction of a leading man of rational, scientific thought, there are hints of a more moralistic and sentimental frame for how this landscape was intended be read. Lettsom, in the concluding sections of his own guidebook, argued that the garden should through beauty "gratify intellect, improve understanding, and inspire the gratitude of a dependent being, whose humility ought to increase with the increase of blessings."[43] This polite sentiment should in turn be encouraged through the contem-

Fig. 4.4. Lettsom, as an expert antiquarian and natural historian, is symbolized by this engraving of an urn, shells, coral, and sculpture, presumably a representation of items held in his museum. From Maurice's *Grove-Hill.* Wellcome Collection, photo by author.

plation of the "arrangement and œconomy of the garden and premises."[44] Alongside a contemplation of what man owes God, Lettsom hoped that Grove Hill would encourage humility and charity in the reader and visitor. The guide then is more than a description of a route around the landscape to take in the various ornamental, botanic, and scientific features, but also an attempt to raise the sensibility of the reader to higher emotions. As Spary has discussed in relation to French natural history, "science" and "sentiment" were not distinct.[45] Similarly, here the organized botanic collection,

Fig. 4.5. The significance of scientific and technological endeavors was stressed by the inclusion of an illustration of the Deptford telegraph system. From Maurice's *Grove Hill*. Wellcome Collection, photo by author.

the observatory, and the statues dedicated to Hygeia, Cupid, and Shakespeare (fig. 4.6), were all set within an ornamental and picturesque frame designed to encourage an emotional response from the reader in relation to their fellow men. Using a definition coined by Ann Jessie Van Sant, we can argue that for Lettsom both "*sentiment/sentimental* and *sensibility*" were related to an immediate moral, scientific, and aesthetic responsiveness that would be elicited by the landscape itself.[46]

As a you might expect of a work distributed solely to friends, there appear to be few published review articles of Lettsom's guidebook, so our understanding of its reception is limited. A rare example published in *The British Critic, A New Review* in 1795 described it as "a model of convenient horticultural arrangement" and suggested that other gentlemen should follow Lettsom's lead, with similar depictions of the "advantages and improvements which his diligence has sought, and his experiments accomplished," as "the benefit would undoubtedly be very great and extensive."[47] A personal response to the work as a model that could be followed was also expressed by the Reverend Plumptre, who was one of Lettsom's regular cor-

Fig. 4.6. Statue of the eminent playwright William Shakespeare, underscoring Lettsom's status as a cultured man. From Maurice's *Grove-Hill*. Wellcome Collection, photo by author.

respondents. On receiving and reading the work, he thanked Lettsom and stated that "as a book, it is beautiful, as a work pleasing, and, should I ever get into a place where I am at all likely to be settled, I shall endeavour to imitate your plans in a humble way."[48] In both these examples, the guide is received less as a way to navigate oneself around the garden as a visitor, but more as a pattern that could be followed to render gardens, as Lettsom hoped, "more agreeable as well as more useful."[49]

A word should be said here about the illustrations that accompany the poem and the guidebook. Some of the same images were used within both works, but the poem in particular made use of imagery that emphasized the classical nature of Grove Hill and Lettsom himself. With depictions of Lettsom dressed in classical robes reading in his library (fig. 4.7) and a print of rural workers gathering the harvest (fig. 4.8), this is a text that places Lettsom firmly within a Virgilian georgic tradition. Overall, the poems and the guide present a particular perspective—of an ordered, scientific, and fashionable rural landscape where nature has been tamed, and overseen by a successful professional man who had power over every ele-

Fig. 4.7. Lettsom depicted as a classical figure surrounded by antiquarian objects, reading in his library, again highlighting his Virgilian and learned identity. From Maurice's *Grove-Hill*. Wellcome Collection, photo by author.

ment of his recreated Eden built on a hill and looking down on London, and therefore the rest of society, below.

Cataloging the Collection

At the end of Lettsom's 1804 edition of *Grove-Hill* there is appended a catalog of trees and plants. This catalog, Lettsom states, was designed to encourage others "to procure cuttings or roots of plants, which he may not possess, as well as to confer similar favours."[50] It therefore acted as both a record

Plate 1. Portrait of Dr. John Coakley Lettsom and his family in their garden at Grove Hill, Camberwell, London. From left to right, the family is thought to be Ann Miers Lettsom (his wife), their children, Mary Ann Lettsom (1775–1802), Edward Lettsom (1781–1821), Pickering Lettsom (1782–1808), John Miers Lettsom (1772–1800), John Coakley Lettsom (himself), and Samuel Fothergill Lettsom (1779–1884). As well as including plant specimens in pots and one of the glasshouses for exotics, the painting also includes references to Lettsom's interests including botanizing through John Miers's handling of a plant specimen, and astronomy is alluded to via the telescope held by Samuel Fothergill. Painted in oil ca. 1786 by an unknown artist. Wellcome Collection,

Plate 2. The only known image of John Hope (left), who is depicted with a gardener, presumably at the Leith Walk garden. Colored etching by John Kay, 1786. Wellcome Collection, CC BY 4.0.

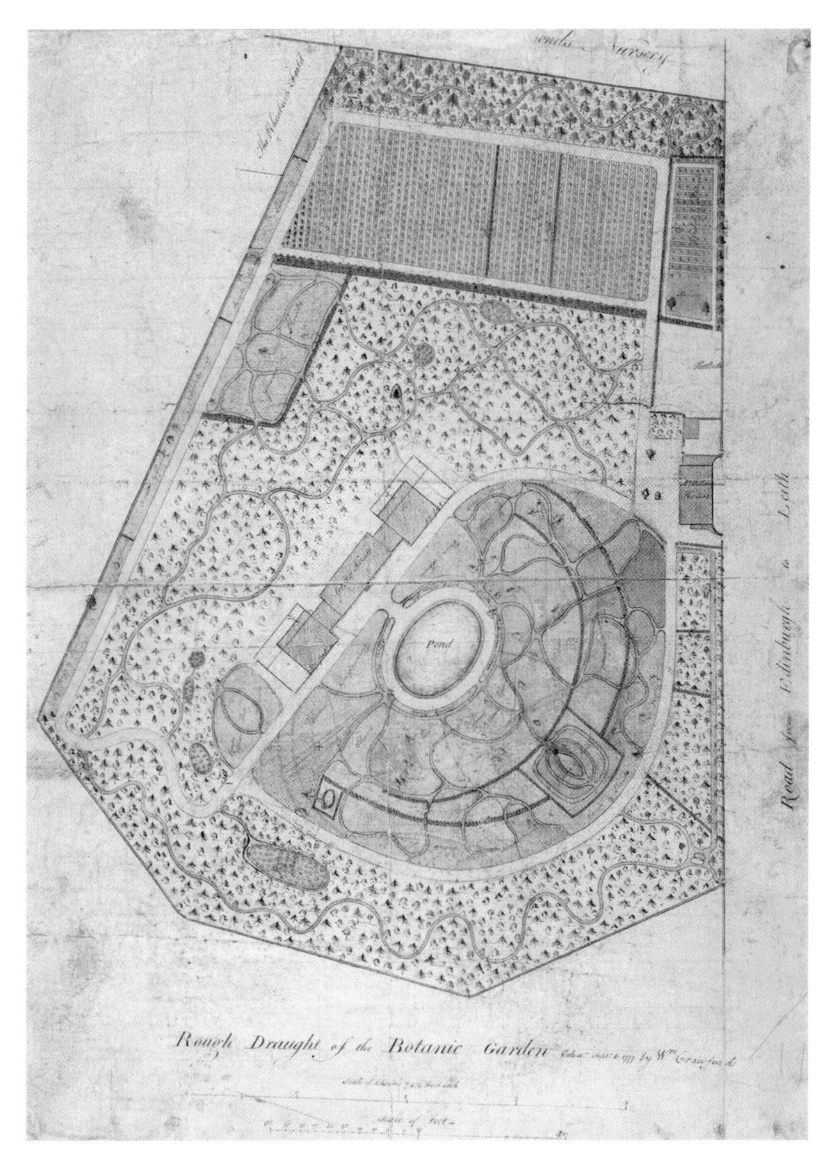

Plate 3. 1777 plan of the Leith Walk garden showing the regimented lines of the Schola Botanica at the top of the image and the winding paths with poppy seed motif to the right of the pond. The botanic cottage, which has now been moved and reconstructed at the current botanic garden, is the red-colored building to the right by the perimeter wall. Reproduced with permission of the Trustees of the Royal Botanic Garden Edinburgh.

Plate 4. This watercolor gives an impression of the use of various specimens, models, and drawings in the teaching of anatomy. *A Lecture at the Hunterian Anatomy School, Great Windmill Street, London,* by Robert Schnebbelie (1830). Wellcome Collection, CC BY 4.0.

Plate 5. The north view of John Coakley Lettsom's house at Grove Hill, Camberwell. Colored engraving by Darton and Harvey after G. Samuel, 1795. Wellcome Collection, CC BY 4.0.

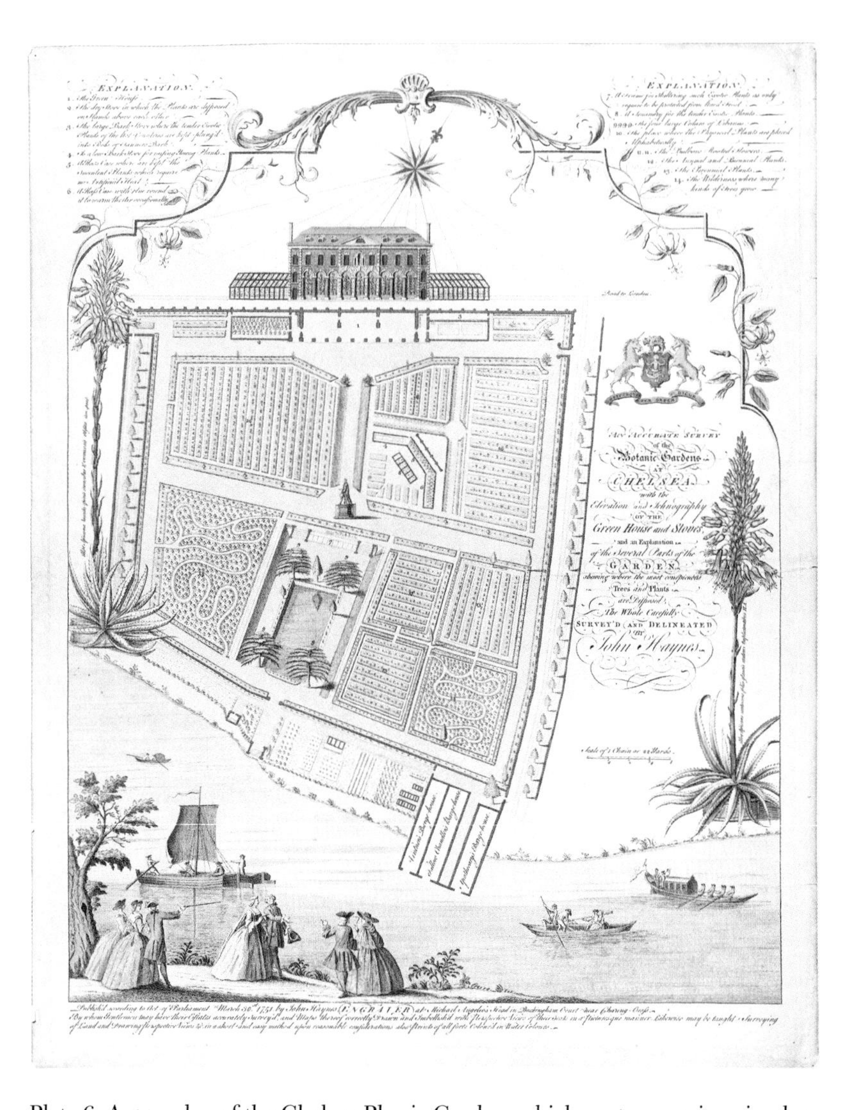

Plate 6. A 1751 plan of the Chelsea Physic Garden, which portrays an imagined arrival of exotic plants on boats along the river Thames, as well as the well-dressed audience on the riverbank, all underlining the fashionable interest in plants. The garden was owned by the City of London's Worshipful Society of Apothecaries and was mainly used to train apothecaries. Engraving by John Haynes, 1751. Wellcome Collection, CC BY 4.0.

Plate 7. View of the wilderness at Kew by William Marlow, 1763, showing William Chamber's magnificent Pagoda as well as the Alhambra and the Mosque. Here the global is clearly presented within what is generally recognized to be the domestic English landscape style. Harris Brisbane Dick Fund, 1925, Metropolitan Museum of Art.

Plate 8. Colored etching by the botanical artist George Ehret, of a magnolia species with flowering stem and labeled floral segments, fruit, and seed, ca. 1737, after himself. Wellcome Collection, CC BY 4.0.

Plate 9. Mark Catesby was the first to publish a natural history of North America. His watercolor, heightened with gum arabic, of an American bullfrog (*Lithobates catesbeianus*), 1722–1726, is likely to be of the same or a similar species as that sent to Fothergill by Bartram from Pennsylvania. RCIN 926025 Royal Collection Trust / © Her Majesty Queen Elizabeth II 2020.

Plate 10. George Edwards's 1757 image of "the small mud tortoise, smelling strong of musk, having a sharp horn poynted tayl from Pensilvania" [*sic*]. At least one of Lettsom's tortoises that lived in his vegetable garden originated from West Chester in Pennsylvania. Wellcome Collection, CC BY 4.0.

Plate 11. The medicinally important foxglove (*Digitalis purpurea*) as illustrated in William Withering's famous account, *An Account of the Foxglove, and Some of Its Medical Uses: With Practical Remarks on Dropsy, and Other Diseases* (Birmingham: Printed by M. Swinney for G. G. J. and J. Robinson, London, 1785). Wellcome Collection, CC BY 4.0.

Plate 12. Thomas Sandby's view of the aviary at Kew gardens from William Chambers, *Plans, Elevations, Sections, and Perspective Views of the Gardens and Building at Kew in Surry*, 1763. Harris Brisbane Dick Fund, 1925, Metropolitan Museum of Art.

Plate 13. View of the New Menagerie at Kew, 1763, again by Thomas Sandby in William Chambers's *Plans, Elevations, Sections, and Perspective Views of the Gardens and Building at Kew in Surry*, 1763. Harris Brisbane Dick Fund, 1925, Metropolitan Museum of Art.

Plate 14. Stipple engraving of the nurseryman and botanist James Lee by Samuel Freeman (1810). Lee is depicted closely observing a specimen with a magnifying glass, while his hat, filled to the brim with flowers, is set to his side. Delineated from an oil portrait by George Garrard. Wellcome Collection, CC BY 4.0.

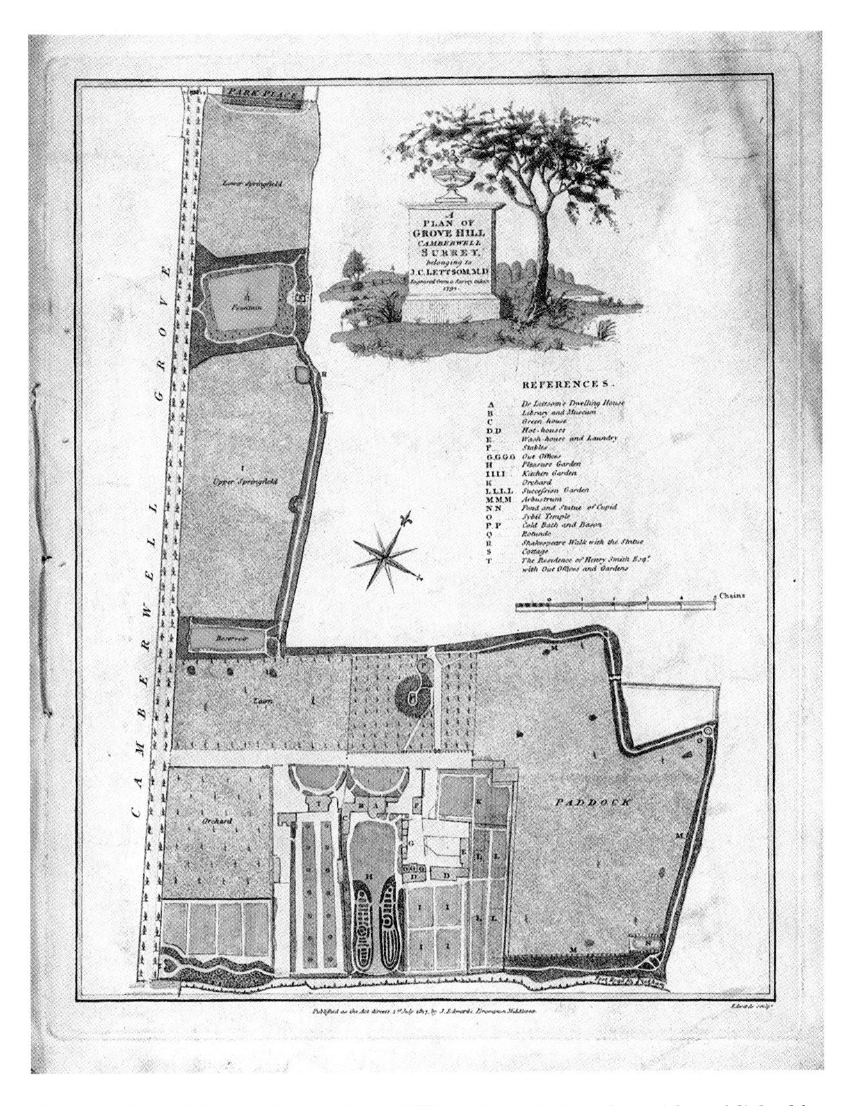

Plate 15. Plan of the gardens at Grove Hill, as included in the guide published by Lettsom with house and adjacent buildings in the central bottom portion. As well as giving a sense of the ornamental nature of the grounds near the house, the plan also depicts a cold bath, paddock, fountain, and orchard. This colored engraving of the plan is by J. Edwards, 1817. Wellcome Collection, CC BY 4.0.

Plate 16. This colored engraving of the tea plant, *Camellia sinensis,* with flowering stem, sectioned leaf, and many floral segments is by John Miller and featured in Lettsom's work *The Natural History of the Tea-Tree* (London, 1772). Wellcome Collection, CC BY 4.0.

Plate 17. James Sowerby's watercolor of William Curtis's botanic garden, when it was based at Lambeth Marsh, ca. 1787. It gives a sense of the ordered beds as well as the ornamental nature of the garden. Private Collection / Photo © Christie's Images / Bridgeman Images.

Plate 18. The south view of Dr. Lettsom's villa at Grove Hill, Camberwell, including a gardener and a fashionably dressed family group, 1817. This view was originally included as an engraving in Lettsom's *Grove-Hill: An Horticultural Sketch,* 1794. Wellcome Collection, CC BY 4.0.

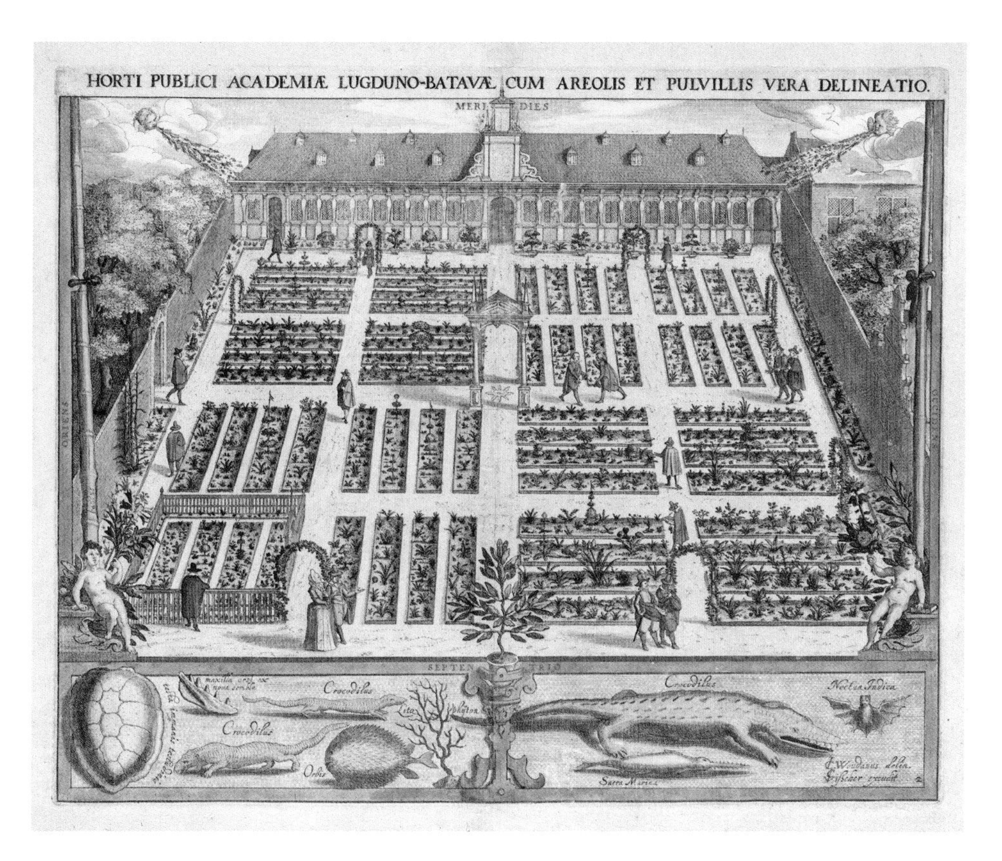

Plate 19. View of the Hortus botanicus of the University of Leiden, with a lower border of natural history objects from the associated collection, demonstrating the close links between the two, 1610. Colored etching and engraving by Willem Isaacsz, from Swanenburg to Jan Cornelisz. van 't Woudt / J. C. Woudanus, published by Claes Jansz. Visscher (II). Noord-Hollands Archief / Voorhelm Schneevoogt NL-HlmNHA_1477_53013257.

Plate 20. Plate of a yellow iris (*Iris variegata*) by James Sowerby for William Curtis's *The Botanical Magazine; or, Flower-Garden Displayed,* vol. 1 (London: W. Curtis, 1787), pl. 16. Wellcome Collection, CC BY 4.0.

Plate 21. The pastoral nature of elite landscapes and the importance of the combination of utility and ornament are illustrated in this hand-colored engraving of King George III's White House at the Royal Gardens at Kew, by James Mason, ca. 1764. Yale Center for British Art, Paul Mellon Collection.

Plate 22. Oil painting of Edward Jenner, attributed to John Raphael Smith, with the rural landscape of Berkeley Castle and a herd of cows in the background, alluding to his crucial work on the cowpox vaccine. Wellcome Collection, CC BY 4.0.

EARL'S COURT HOUSE (FORMERLY JOHN HUNTER'S). (*See page* 161.)

Plate 23. Wood engraving of John Hunter's home at Earl's Court, date unknown but presumably made after his death. Wellcome Library no. 561251i. Wellcome Collection, CC BY 4.0.

Plate 24. Watercolor of the mound at Earl's Court, with a pencil date of 1783, showing the mound as a pastoral feature from the Hunter family album. From the Archives of the Royal College of Surgeons of England.

Plate 25. Gardens were used for lavish parties throughout the eighteenth century. This French oil on canvas of an imagined *fête champêtre* set within the landscape dates from around 1730 and is by Jean-Baptiste Joseph Pater. Samuel H. Kress Collection, Courtesy National Gallery of Art, Washington.

Plate 26. Actor and playwright David Garrick and his wife by his Temple to Shakespeare at his Hampton villa on the outskirts of London, ca. 1762. Johan Joseph Zoffany, oil on canvas. Yale Center for British Art, Paul Mellon Collection.

Plate 27. The fountain and cottage at Grove Hill, with a party enjoying the sociable activity of boating on the lake. Colored engraving after G. Samuel, 1795. Wellcome Collection, CC BY 4.0.

Plate 28. John Spyers's watercolor of the King's Observatory at Richmond, which was designed by William Chambers and built for King George III. British Library, Maps K.Top.41.16.r.

Plate 29. The fashionable activity of stargazing as satirically illustrated by Thomas Rowlandson, at the Oxford Observatory, 1810. The Elisha Whittelsey Collection, The Elisha Whittelsey Fund, 1959, Metropolitan Museum of Art.

Plate 30. Photograph of a cork model of the Colosseum in Rome, made by Richard Du Bourg, London, ca. 1775. Museums Victoria Collections, CC BY 4.0.

Plate 31. The relocated and restored botanic cottage at the Royal Botanic Garden Edinburgh, where it is having a second life as an education and community hub, set within an appropriately scientific demonstration garden. Photo by author, 2016.

Plate 32. A sign placed in a plant border within the Oxford Botanic Garden, encouraging visitors to sensorially interact with the plants by smelling their leaves rather than just observing them from a distance. Photo by author, 2018.

Fig. 4.8. The arcadian and georgic nature of Lettsom's country retreat is emphasized by this illustration of agricultural workers bringing in the harvest, with the Thames and the City of London in the background. It also represents the situation of Grove Hill on the edges of the city. From Maurice's *Grove-Hill*. Wellcome Collection, photo by author.

and advertisement of Lettsom's activity and also as a way of facilitating the botanic networks, which, as we have seen, were so integral to plant collecting in this period. Catalogs then, alongside their inherent commercial nature when produced by nurserymen in particular, were another way of communicating via print regarding the status of a plant collection and encouraging future exchange of specimens. They could also be created to advertise the polite, botanic status of their owner, as well as to offer distant readers an imagined tour of the specimens to be found growing in the garden.

From their inception university botanic gardens produced lists of plants detailing what was growing in their collections, as befitted their status as new scientific institutions.[51] Similar to those produced by private garden owners like Lettsom, these catalogs were invaluable to serious botanists who relied on the crucial exchange of seeds, bulbs, and plants to build their collections.[52] Originally produced by hand, such lists were soon published so that they could be disseminated widely, initially nationally, but later to a global community of botanists. Printed catalogs, like guidebooks and other ephemera, were a growing phenomenon during the eighteenth century. Just as botanic gardens did, nurserymen produced catalogs to advertise the plants available for purchase.[53] Catalogs and lists created by private owners can then perhaps be seen as analogous, as they also advertised the owner's botanic collection far beyond their immediate geographic and specialist networks and facilitated the development of wider networks of interested peers as well as more extensive collections of plant material.

Such lists could also act as memorials. On Fothergill's death in 1780, Lettsom prepared a list of the plants currently growing in the late doctor's garden, *The Hortus Uptonensis, or a Catalogue of Stove and Greenhouse Plants.*[54] This may have been rooted in a desire to collate a permanent record of an extensive and nationally important collection, which was also a manifestation of the owner's loves and interests. In the note inserted before a reproduction of the catalog in Lettsom's *The Works of John Fothergill,* he states that the catalog was organized alphabetically in order to be "most familiar to the generality of readers, and more convenient for those who may be fond of horticulture," suggesting that it was intended to be of practical value.[55] Lettsom himself obtained two thousand of Fothergill's plants after the auction of 1781, which he transplanted to Grove Hill.[56]

In the same note, Lettsom argued that he had created the catalog to share expert advice on plant collecting and growing. For example, the catalog itself includes "the situation each requires, by the letters S and G; the former signifying the stove or hot-house, and the latter the green-house."[57] He also states that he included names in the Malay language where he thought it would be useful to enable "the inquisitive traveller to discover more readily and certainly the object of his enquiries."[58] This is an interesting case in which the importance of local and indigenous knowledge in the art of botanic collecting is recognized—in this case in Indonesia and the

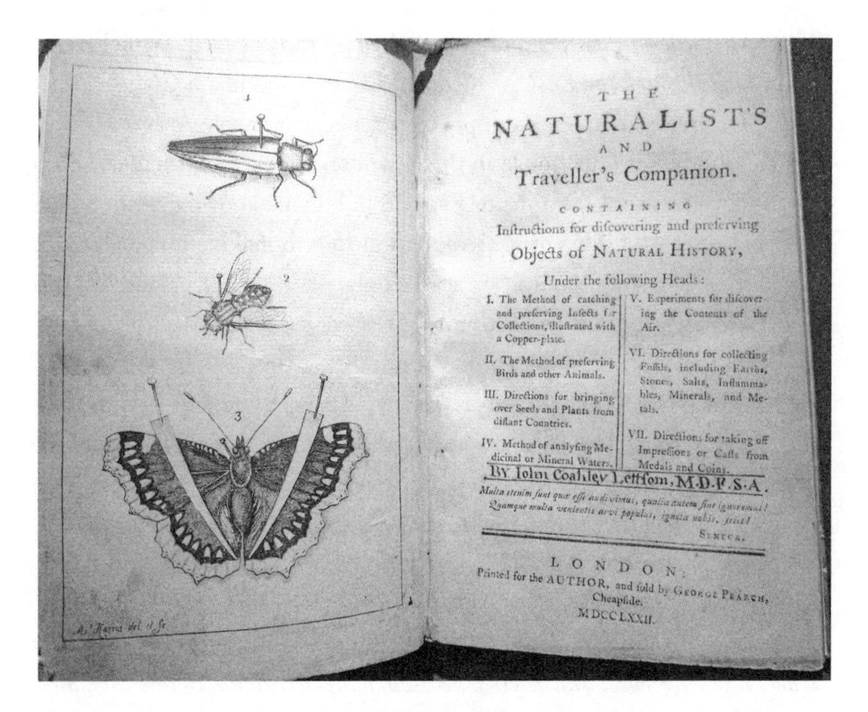

THE
NATURALIST'S
AND
Traveller's Companion.
CONTAINING
Inſtructions for diſcovering and preſerving
Objects of NATURAL HISTORY,
Under the following Heads:

I. The Method of catching and preſerving Inſects for Collections, illuſtrated with a Copper-plate.

II. The Method of preſerving Birds and other Animals.

III. Directions for bringing over Seeds and Plants from diſtant Countries.

IV. Method of analyſing Medicinal or Mineral Waters.

V. Experiments for diſcovering the Contents of the Air.

VI. Directions for collecting Foſſils, including Earths, Stones, Salts, Inflammables, Minerals, and Metals.

VII. Directions for taking off Impreſſions or Caſts from Medals and Coins.

BY John Coakley Lettsom, M.D. F.S.A.

Multa etenim ſunt quæ eſſe audivimus, qualia autem ſint ignoramus! Quamque multa venientis ævi populus, ignota nobis, ſciet!

SENECA.

LONDON:
Printed for the AUTHOR, and ſold by GEORGE PEARCH, Cheapſide.
MDCCLXXII.

Fig. 4.9. Title page of Lettsom's *The Naturalist's and Traveller's Companion,* including an image of the local Camberwell butterfly (London: Charles Dilly, 1772). Wellcome Collection, CC BY 4.0.

surrounding area—although the work itself is still aimed at the European traveler. The descriptions of how to successfully transport seeds and plants from foreign climes within the catalog have parallels with Lettsom's earlier 1772 work (fig. 4.9), *The Naturalist's and Traveller's Companion, Containing Instructions for Collecting & Preserving Objects of Natural History and for Promoting Inquiries after Human Knowledge in General.*[59] Plant catalogs, then, offer the reader more than just an armchair journey through descriptions of the plant material grown in the garden. They can also offer advice and knowledge regarding how to grow plants, their native names, and how to transport botanic seeds and specimens.

Some catalogs also raise questions regarding their accuracy as a snapshot of a living collection and how broad a scope they may have taken regarding what was included. Dr. William Beeston Coyte inherited and developed a garden in the center of Ipswich.[60] As early as 1722, Daniel Defoe in his *Tour through the Eastern Counties* wrote that William Coyte's great-uncle "Dr. Beeston, an eminent physician, began a few years ago a physic garden adjoining to his house in this town; and as he is particularly curious, and, as I was told, exquisitely skilled in botanic knowledge, so he has been not only very diligent, but successful too, in making a collection of rare and exotic plants, such as are scarce to be equalled in England."[61] This was clearly an important collection, as Beeston was also presenting exotics as gifts to the new university garden at Cambridge.[62]

In the 1760s, William Coyte returned to Ipswich and took over the running of the garden. This garden is featured on Pennington's 1778 map of the city and is recorded by Coyte in a catalog with the extremely long title, even by eighteenth-century standards: *Hortus Botanicus Gippovicensis; or, a Systematical Enumeration of the Plants Cultivated in Dr. Coyte's Botanic Garden at Ipswich, in the County of Suffolk, Also, Their Essential Generic Characters—English Names—The Natives of Britain Particularised—The Exotics Where Best Preserved, and Their Duration; with Occasional Botanical Observations, to Which Is Added, an Investigation of the Natural Produce of Some Grass-Lands in High Suffolk.* Like *The Hortus Uptonensis,* the catalog represents more than a list of plants, as it also includes notes on how to preserve exotic specimens alongside the results of experiments with domestic grasses. This, like many printed botanic works from this period, underlines both the global and local nature of horticultural science. Knowing how to grow new exotic species in hothouses was just as important as recognizing and understanding the growth habits of local plant life, as both had economic value.

The 158-page booklet contains an extensive plant list, including 120 hardy tree species, which has led Blatchly and James to question whether all of these could have been grown in a relatively small town garden.[63] They conclude that although the garden was potentially large enough to hold the number of plants listed, there would have been limitations, as the diversity of plants implied a very varied range of habitats.[64] One example they provide is the inclusion of a number of seaweeds but with no indication of a pond or any other suitable watery habitat for their cultivation. They argue that these may have instead formed part of a dried rather than a living collection. Similarly, they suggest that there is no indication of temporality, so that plants grown earlier in the century may have been included alongside those currently thriving in the space.

Catalogs, with their temporal and spatial slippage, then, represented more than simple plant lists of living specimens and could have been both published and read for different reasons in this period. Coyte himself was a member of the prestigious Linnean Society, and the printed work may have been an attempt to cement his status as an eminent botanist. In 1789 he wrote to the then president of the society, James Edward Smith, exclaiming that he was "eagerly pursuing my inclination for a large collection and every requisition gives me much pleasure."[65] He also exchanged seeds with Smith and James Sowerby as well as collected them from the local area on herborizing trips. His garden, like the others in this book, formed part of the broader network of botanic collections and collectors, often found coalescing around larger urban centers, such as London, and the printed catalog would have given him recognition farther afield.

Labeling Specimens

The eighteenth-century scientific need to create order out of chaos can also be seen in the use of plant labels. This was particularly relevant for the university botanic garden, where trees and shrubs without labels had little value as scientific or teaching specimens. A lack of clear organization risked the designated and ordered space becoming viewed as a pleasure, rather than a scientific, garden.[66] Private botanic garden owners also stressed their personal scientific sensibility through the use of labels to elevate their collections from mere trifles for leisured appreciation.

For visitors, one of the key methods of reading a botanic garden was (and still is) via plant labels placed close to each physical specimen. In his *Proposals for Opening by Subscription a Botanic Garden to be Called the London Botanic Garden,* Curtis describes how he had applied Linnaeus's new taxonomic language to his collection at the London Botanic Garden: "To each plan in the garden, is affixed its generic and trivial name, according to LINNAEUS: and that none may lose the advantage of acquiring a knowledge of plants from a nonacquaintance with the Latin, the English names also are added."[67] The use of English alongside Latin emphasises the educational role of such gardens in relation to new botanical methods of organization. This approach was also taken in Lettsom's domestic garden, as we have seen. As well as spurring on botanic activity, as discussed in chapter 1, Linnaean classification aided in the simplified labeling of plants, as they could now be given names directly on their labels rather than simply being numbered and then read via an associated handlist, as at Montpellier and elsewhere.[68] This simplified approach to labeling also meant that there was an increase in popularity among the educated public, who could presumably more readily "read" these plant collections.

Curtis also included a color-coded system of labels, with plants marked yellow for those useful in medicine, blue for food, black for poison, red for dying, and green for agricultural purposes.[69] The educational purposes of the garden in relation to medicinal knowledge can be seen too in Curtis's aim to include all the physical plants listed in Lewis's *New Dispensatory* as well as other more novel introductions.[70] He stated that he included labels with both the common and Linnaean names as well as a reference to the page and article of relevant entries in the *Dispensatory*. To make the most of this he recommended that students remove the materia medica part of the *Dispensatory* so that it might be a more portable handlist.[71] Similarly, Hope explained to students that "the officinal plants . . . are all numbered and you have catalogues of them in your hands."[72]

This illustrates the close relationship of textual knowledge and physical collections and how visitors and students were expected to use the written word as they navigated the collection. The education of the botanist using both the physical plant material and a text reflected the close relationship between written works and the training necessary to become an expert in close observation. For example, works such as Linnaeus's *Philosophia Botanica* were not just guides to his particular method of classification but also helped readers become expert observers, using the combination of image and text to train the eye.[73] By close scrutiny of the text, the plates, and the specimens, a student could train themselves in the observational techniques needed for botanical science.

Labels could also reveal more than just the scientific names of plants. In the new Dublin botanic garden, established in Glasnevin in 1796, each plant had fixed alongside it a painted mark that corresponded to the garden catalog.[74] In this case, though, the labels also demonstrated nationalistic sentiments.[75] For example, the letter "N" was painted on the back of labels to indicate native species of plant. The botanic garden at Glasnevin was definitively agricultural rather than medically motivated, partly due to the involvement of the Royal Dublin Society, which had been established in 1731 to improve the poor economic condition of the country by promoting agriculture, arts, industry, and science in Ireland. This was also in line with its key instigator, John Foster, whose personal interests were more allied to national economic success than purely scientific concerns.[76] As the geographer Johnson argues, based on "the significance of agriculture to Ireland's domestic economy, and the presence of a land-owning elite among the members of the RDS, it is not surprising the status awarded to the agricultural dimension of the garden."[77]

Finola O'Kane, in her work on the Dublin garden, also demonstrates how nationalism was tied into the agricultural aims:

> The desire for a public garden lay in its instigator's wish to educate both their peers and the common man and to disseminate as widely as possible the economic and social benefit wrought by improvement; improvement being part and parcel of both colonization and the rising capital economy. This gave the new foundation distinctly nationalist and competitive overtones.[78]

This reflected the impetus seen behind the calls during this period for new botanic gardens based on national, regional, and/or civic competitiveness and local pride. This fostering of local identity through botanic collecting also reflected the production of texts describing local flora, such as Curtis's *Flora Londinensis* and Lightfoot's *Flora Scotia,* for which Hope wrote the preface, as well as many other regional floras. This again demonstrates the close links between the physical plant collections and the written texts.

A Living Library

The botanic garden, by having species of plants that were easily identifiable by their labels and cataloged, was, like a scientific museum collection, arranged to be read in a similar manner to a library.[79] It can, therefore, be viewed as a living library, which once arranged scientifically could then be read. It is perhaps unsurprising that many of the gardens with botanic collections also had their own libraries physically located either within them or set in houses sited on their peripheries.

Country houses and botanic gardens obtained libraries of works discussing plants and horticultural methods, so that the physical nature of the plant material could be identified, understood, and given meaning through the written word. This interconnection between libraries and gardens can also be seen in the creation of a specific building type, known as the garden or summer library.[80] One example is the domed rotunda, known as the Mussenden Temple, built in the grounds of Downhill House in what is now Northern Ireland. Like the observatory at Grove Hill, explored in the following chapter, this library was also based on the Temple of Vesta at Tivoli and certainly contained a library in the 1800s.[81] The living and the textual library were then intertwined and physically interconnected, and both were necessary for scientific understanding.

Botanic specimens also moved from the wild and the garden into the library as they were cut, dried, and preserved in herbaria. On the sale of Lettsom's library in 1811, it was noted that it included sixty-two volumes of his own herbarium, or hortus siccus.[82] This highlights the direct link between the paper volumes and the living collections. This interconnection between specimens, botanists, and the annotated document that formed the pages of herbaria is constantly transformed as notations are added as knowledge changes or other botanists add their own mark. The herbarium sheet is part preserved object, part botanic annotations.[83] So the library includes active material whose value is added to over time and is used to help read other texts as well as the living library in the garden.

In addition to including herbaria, libraries included essential reference books for botanists. Curtis's London Botanic Garden incorporated a library for the use of subscribers from its very inception. According to Edwards's *Companion,* Curtis already had "a considerable library and an extensive collection of drawings in natural history" when the London Botanic Garden was in its first location in Lambeth.[84] William Noblett argues that this library, described by Curtis as "chiefly in Natural History, Medicine and Agriculture," was "the very first single subject library to be part of a specialized educational establishment, and the first subscription library to be given over to a single scientific subject."[85] This suggests that not only were library and garden collections intertwined, but they were also co-developing as cutting-edge approaches to the organization and dissemination of knowledge, whether in written or living format. This state-of-the-art approach also mirrored that taken by the most advanced scientific research institutions in Europe and North America, which were not only investing in the top scientific instruments, such as observatories, laboratories, and botanic gardens, but also locating libraries at the heart of them.[86]

In the 1800s, by which time the London Botanic Garden had moved to Brompton and was being run as a joint venture by Curtis and Salisbury, a published and comprehensive catalog offers us an insight into the later purposes of the garden, including the extent of the library, which they advertised to subscribers as including

> studies of taxonomy and practical botany by Linnaeus, Adamson, Martyn, Withering, and others, natural historical works, dissertations on medical botany, accounts of voyages and international exploration, national and regional botanical surveys such as Charles Deering's *Catalogue of Plants about Nottingham* (1738) and John Lightfoot's *Flora Scotica* (1777) and White's *Natural History And Antiquities of Selborne* (1789), journals from relevant academies and scientific associations and books on gardening, horticulture, agriculture, and arboriculture.[87]

From this, it is evident that the library not only offered tomes useful for reading the collection immediately outside the reading space, but also contained works that could be used to place it within its wider national and global context, again connecting the garden and the interests of domestic botany to the broader economic and scientific aims of the empire. The roots of the botanic collection again reached far beyond the confines of the perimeter of the garden.

The location of the library at Brompton, with a separate building placed between the head gardener's house and an aviary, also links us back to those country-house summer libraries situated in the landscape.[88] The library was connected to the natural world beyond its location within the garden, as the latticework allowed the reader from within the library to observe and learn the habits of the native birds (possibly those kept within the nearby aviary). A table in the center, covered in a green tablecloth, also formed the backdrop for the investigation of flowers and other plant specimens brought in from the garden for closer inspection.[89] In Thornton's sketch of Curtis he described how at Brompton "every work of importance in this branch of science, is to be met with here: and the student may fancy himself a thousand miles from London, only occasionally interrupted by the melody of songsters in the aviary."[90] The living and the written formed the sum of knowledge in a recreation of a rural Elysium.

Physical and intellectual interconnections were also made with natural history collections. From at least the end of the sixteenth century, the botanic garden at Pisa had an adjoining gallery, which John Evelyn, on a visit in 1644, recorded was "furnish'd with natural rarieties, stones, minerals, shells, dry'd Animals, skeletons &c as is hardly to be seen in Italy."[91] This juxtaposition between collections can also be seen in the 1610 illustration of the Leiden botanic garden (plate 19), where a range of natural history specimens were depicted around the margins of an illustration of the botanic garden, thereby placing the two in dialogue. Similarly, the ornamental but highly fashionable grottoes created in many gardens, including by Mrs. Delany herself for a range of clients, with their decorative shells, crystals, and other ornamentation, can also be read as a natural history cabinet.[92] All represented the growth in scientific knowledge of the natural world and the interweaving of different forms of collecting and understanding, between the interior and exterior, as well as between the domestic and institutional and the wider world.

Back at Grove Hill, it is no surprise that we discover that Lettsom also had an extensive library adjoining the garden. The printer, author, and antiquarian John Nichols remembered how "Dr. Lettsom's Library was ample, and contained such a collection of books in all languages, and on all sciences, as few private gentlemen possessed."[93] When Lettsom was eventually forced to sell the greater part of his library in 1811, it took eight days, and the sales catalog included 1,347 volumes, demonstrating its sheer extensiveness.[94] Its relationship to both the garden and Lettsom's other collections was evident in its physical location. According to his 1794 sketch, "The library opens by a glass door into the garden through the greenhouse; and by another door into the museum or repository for natural history and other curiosities. The marble chimney-piece in this room is carved in shells, equal to fine natural specimens."[95] As noted above, this highlights the common relationship between botanic and other collections, whether of books, antiquities, or works of art. Similarly, *Jackson's Oxford Journal* in 1773 reported a notice outlining "the Museum of Natural Curiosities, now forming at the Botanic Garden by the ingenious and indefatigable Professor Martin."[96] This suggests that collections, including botanical ones, need to be considered in relation to one another as well as on their own terms.

Bleichmar argues that during this period the relationship between various collecting practices and activities was accompanied by a rise in the concern with visual expertise. As with the use of written and illustrated texts to train the eye, she contends that "what brought things like gardens, shells, and paintings together was a concern with visual expertise, with outlining and deploying practices of specialized diagnostic looking; and, second, that this interest in the enskillment of vision arose from acts of collecting and display."[97] Although looking is clearly central to many of these activities and ways of knowing, in relation to plants in particular this was not necessarily just a visual endeavor but also something that engaged the other senses. All worked together in a concerted attempt to try to make sense of the natural world.

Published texts also highlighted the knowledge and interests of owners. William Curtis, for example, produced a variety of texts including his volume on grasses (fig. 4.10), the *Flora Londinensis,* and his *Botanical*

Fig 4.10. Plate 3 of smooth-stalked meadow grass (*Poa pratensis*), from William Curtis's *Practical Observations on the British Grasses Best Adapted to the Laying Down, or Improving of Meadows and Pastures: To Which Is Added, an Enumeration of the British Grasses,* 2nd ed. (London: Printed by Couchman and Fry, and published by the author, at No. 3, St. George's-Crescent, near the Obelisk, Black-Friars-Road, 1790). Wellcome Collection, CC BY 4.0.

Magazine, and Lettsom also produced many and varied pamphlets, including one on the cultivation and uses of the mangel-wurzel. Similarly, nurserymen also produced scholarly works. John Lee of the Vineyard Nursery produced an *Introduction to Botany* in 1760 which was predominately a translation of Linnaeus's *Philosophia Botanica.* This was a popular work and conferred upon Lee the status of an expert botanist as well as a successful nurseryman. Like other botanists, Lee's expertise was not confined to botany. According to his contemporary the physician and botanist Robert John Thornton, "He was also adept in entymology, conchology and natural history in general of which he made a most superb collection."[98] Lee was, therefore, both a gentleman interested in polite knowledge and someone who exploited this knowledge, quite reasonably, for commercial gain. Similarly, the proliferation of early-eighteenth-century nurseries provided the public with accessible gardens that both encouraged and aided the new interest in gardening as a scientific pursuit as well as gave them the opportunity to purchase these new specimens for their own gardens.[99]

Thornton's own opinion of Lee's nursery was that "although the great exertions made to extend the Royal Garden, at Kew, and large sums expended, made that the chief repository of new and rare plants, still Mr Lee's Nursery at Hammersmith, took, at any rate, the second lead; and the two imperceptibly as it were, enriched our gardens and extended the Science of Botany."[100] The spaces themselves also encouraged both the dissemination of scientific knowledge and opportunities for sociability, much in the same way as the meeting rooms at the Royal Society or the popular haunts of coffee shops.[101] This enables us to consider the institutional alongside the private and the commercial, and also to consider the different ways in which knowledge was created and disseminated within these spaces and even beyond, through published texts.

The relationship between textual productions and physical gardens could work both ways. For example, readers of Curtis's magazine could, from the comfort of their own armchairs, imagine walking around the London Botanic Garden via the plates and descriptions (plate 20). As Elliott describes, "As the alternative title of 'Flower-garden Displayed' suggests, the *Botanical Magazine* proceeded an imaginary walk through a botanic garden and Curtis emphasises the fact that the hand-coloured engravings representing 'natural colours' were based upon living specimens, especially those in his botanic garden."[102] As with the labels in the ground, these tools enabled readers to engage with the material productions of the garden. In *The Botanical Magazine,* text was used to "delineate the name, class, order, generic and specific character according to the Linnaean system so that 'ladies, gentlemen and gardeners' would become 'scientifically acquainted' with the plants they cultivated."[103] Gardens then could be both read from home via a text, such as Lettsom's sketch or Curtis's *The Botanical Magazine,* or the texts themselves could be used within the garden space to literally help users of the collections to navigate their way around the space and to find meaning within the living library.

Artistic representations, as well as documenting flourishing plants, could also provide a record of plants that did not flower successfully or only lived for a short period of time. Fothergill, for example, employed top botanical artists such as Georg Ehret, John Abbot, and J. S. Miller to create lasting records of his sometimes fleeting botanical collection. Banks recorded how "that science might not suffer a loss when a plant he had cultivated should die, he liberally paid the best artists to draw the new ones as they came into perfection; and so numerous were they that he found it needful to employ three or four artists in order to keep pace with their increase."[104] Around two thousand of these paintings commissioned by Fothergill were later obtained by Catherine the Great, the empress of Russia, and were presumably incorporated into her own extensive collection.[105] In this way the garden was also given an afterlife through artistic representations as well as via written accounts and dried herbaria. This afterlife of specimens also resonates with natural history museum collections wherein objects that were once living were given a second life as items for scientific inquiry as well as popular entertainment.

Illiterate Garden Visitors

The written texts, labels, and libraries were obviously aimed at a literate, educated audience. However, there were other oral methods of disseminating knowledge to the less literate visitor. In the prospectus for Dublin's botanic garden of around 1795, a plan was depicted that outlined various areas, including those devoted to different agricultural activities.[106] The prospectus describes two of these areas as follows:

> These hay and cattle gardens are proposed for the instruction of the practical husbandman, he will there see every Plant, Shrub and Weed which grows in Ireland; he will see at once, what are useful, what otherwise, for each Animal; he will learn how to weed his meadows and pastures, how to select the Hay Seeds which should be sown together and what Weeds on his Ditches or Tillage grounds he should be most anxious to prevent seeding.[107]

The intended audience included the illiterate husbandman, as the prospectus states that "the most illiterate Man is capable of Instruction from these, by being told what is the Description of the Division he looks at."[108] The prospectus also notes that "there shall be a professor who shall give Lectures on Botany in general; and also separate Lectures on the Cattle and Hay gardens for the Instruction of common farmers, their Servants, or Labouring Men, all of whom are to be admitted to the Lectures gratis."[109] This has resonances with the London Botanic Garden, where Curtis gave lectures to subscribers, and of course Hope gave botanic lectures for students, although these were both designed for more literate and learned audiences. In Dublin, there was evidently an expectation that the garden audience would be more mixed in nature and that those from the laboring classes could attend such sessions for free. The verbal lectures here would also stand as replacements for the important guides and associated texts that could be read alongside the specimens by more literate audiences.

Similarly, the long history of lecturing within the botanic garden was continued within the new subscription and public botanic gardens. Curtis, for example, gave courses of botanical lectures within the garden and its associated buildings, as well as herborizing trips, on which conversation could replace the written word. Gardeners would also have been available on-site to offer verbal knowledge and advice regarding various plants. On a to-do list compiled by Hope at the Leith Walk garden in 1783, he placed the issue of the head gardener's wages at number one as well as describing the other roles that could result in financial payments to the gardener, including selling plants and showing the garden.[110] As we have seen, being shown the garden was a common occurrence, and surely one that would have had added value for anyone unable to read all the associated labels and texts. Although people were expected to pay the gardener for these expert tours—so they were only available to those who could afford it—it does highlight the variety of different methods by which gardens were read and meaning was constructed.

By considering the mechanisms available to garden visitors, whether scientific botanists or armchair promenaders, we can gain an insight into how owners intended their botanic collections to be understood. It is much harder to find out how visitors themselves interacted with the specimens and the designed spaces, but there were evidently a variety of methods by which one could interpret and find meaning in them. It is also clear from these examples that scientific and educational practices were being advanced through gardens in the same period as well as the better-known classical allusions of the eighteenth-century landscape garden.

FIVE

For Dulce *and* Utile

The Garden as Both Ornament and Farm

IN THE SUMMER OF 1786, Sir Richard Jebb obtained the seeds of a plant perhaps best described as a cross between a turnip and a beetroot, grown in Europe but new to Britain, known as the mangel-wurzel, or, in France, *racine de disette.*[1] Jebb presented a number of these seeds to the Society for the Encouragement of Arts, Manufactures and Commerce (better known today as the Royal Society of Arts), and from there the committee sent them out to members who expressed interest in trialing these new plants.[2] Among this number was Lettsom, who "threw the seed into some light earth placed in a hot-house."[3] In that location he reported that they did well and so he transplanted them into the open air.

In a similar approach to the horticultural experimentation with other more exotic plants, Pettigrew describes how once they had reached maturity, "the plants were examined, and subjected to experiment."[4] As well as experimenting with the seeds, Lettsom translated and edited a French text, *An Account of the Culture and Use of the Mangel Wurzel, or Root of Scarcity* (fig. 5.1), written by L'Abbé de Commerell, within which he also outlined his own findings at Camberwell and argued that this vegetable would be a valuable addition as it was such a productive crop. This he stated was based on his own findings that "a square yard of ground, planted with mangel-wurzel, will yield no less than 50 pounds in weight of salutary food."[5] Lettsom, then, was conducting agricultural experiments at Grove Hill while also developing his scientific botanic collection.

This use as an experimental space for agricultural as well as ornamental plants blurs the boundaries of farm and garden and suggests that the sci-

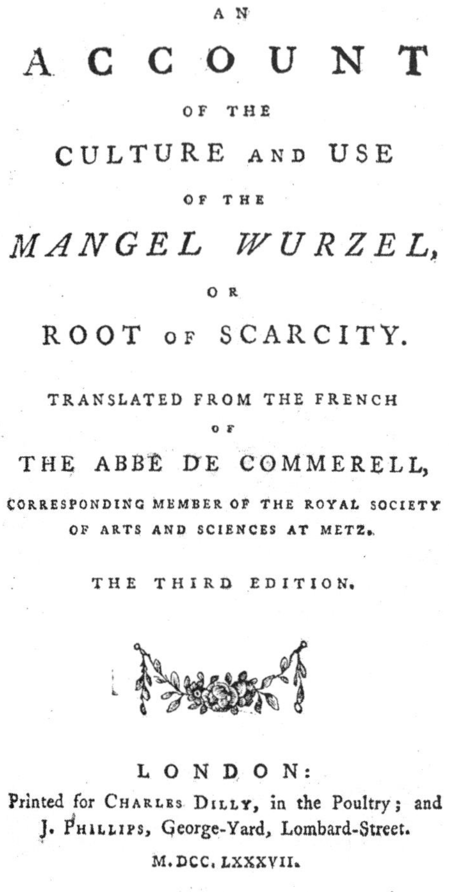

AN

ACCOUNT

OF THE

CULTURE AND USE

OF THE

MANGEL WURZEL,

OR

ROOT OF SCARCITY.

TRANSLATED FROM THE FRENCH

OF

THE ABBÉ DE COMMERELL,

CORRESPONDING MEMBER OF THE ROYAL SOCIETY
OF ARTS AND SCIENCES AT METZ.

THE THIRD EDITION.

LONDON:
Printed for CHARLES DILLY, in the Poultry; and
J. PHILLIPS, George-Yard, Lombard-Street.
M.DCC.LXXXVII.

Fig. 5.1. Frontispiece of Lettsom's *An Account of the Culture and Use of the Mangel Wurzel, or Root of Scarcity,* 3rd ed. (London: Charles Dilly, 1787). Wellcome Collection, photo by author.

entific nature of the space was not confined solely to botanical experimentation. As Timothy Raylor has already argued in his work on early modern cultivation, it is essential to consider the cultural contexts in which agriculture and science operate, and a focus on the garden allows us to consider the interrelationships between the various forms of scientific, artistic, and social activity facilitated by horticultural practice.[6]

In 1773, Richard Weston posed the question, "What can be more amusing than experimental agriculture?"[7] He argued that the fullest amusement would come from

> trying the cultivation of the newly-discovered vegetables, and all the modes of raising the old ones; bringing the earth to the finest pitch of fertility, and raising plants, infinitely more vigorous and beautiful than any in the common tillage; using the variety of new machines perpetually invented and observing their effects; in a small extent of ground, seeing the growth of an infinite variety of vegetables.[8]

This interest in experimentation would have other benefits beyond the purely scientific, which corresponded with eighteenth-century ideas of a polite sensibility. Weston finishes his call to agricultural arms by stating that this approach "gives the most beautiful colouring to every object around, and pleases the refined imagination, with the enchanting prospect of all the elegance of nature."[9] Lettsom, with his various agricultural experiments at Grove Hill, was therefore part of this wider cultural interest in experimentation for both national benefit, in terms of economic botany, as well as the development of an aesthetic appreciation of the spectacle of agricultural trials. This blending of the garden, park, and farm is often overlooked by historians when considering the concerns of the eighteenth-century landowner. However, strong connections were made between beauty and utility which influenced the experience and perception of landscape during this period.[10] This idealized creation of a rural landscape within the boundary walls was an essential aspect of eighteenth-century landscape design.

This can also be seen in the work of John Lawrence, who in 1801 argued that the home farm along with the landscape park should be utilized as a "theatre for the display of *all* the notable varieties of experimental husbandry."[11] It is possible to assume from this that the spectacle of agricultural experiment also formed an element of the Georgian landscape-visiting experience. This is evident in Arthur Young's note regarding the Marchioness of Salisbury's experimental garden at Hatfield, in his *General View of the Agriculture of Hertfordshire* (1804), which described how

> the cleanness of the crops their flourishing luxuriance the general aspect of the whole are truly pleasing. I could not, however but regret that a register had not been kept of every crop, the expense, produce and consumption per acre; this field would then not have yielded pleasure only but an ample harvest of agricultural knowledge and with a few variations easy to have devised would have produced a fund of important conclusions.[12]

Despite the good condition Young found the crops in, it is clear that to him the marchioness's interest appeared to have reflected agriculture as a fashionable and pleasurable pursuit, rather than serious scientific interest. There was certainly a courtly interest in botany and agricultural improvements under George III (plate 21).[13] That setter of trends, the king, not only erected his own model farm adjacent to the Richmond Gardens, but also wrote letters under a pseudonym to Young describing agricultural experiments taking place at Windsor during the 1780s.[14] There is also a sense in Young's account of a tension between rigorous rural experimentation and the growing magnet of urban sociability, as he finishes his account by recording that the "thought had great merit and I cordially wish the field to be so productive of pleasure to its Mistress as to give charms to the country sufficient to rival the great foe to experiment—London."[15]

Lettsom may not have had such an extensive estate, but professional men and wealthy farmers could use what Spooner has described as "a shared grammar of design" in order to create similar, if smaller scale, designed landscapes.[16] This is clearly the case at Grove Hill, as it included many garden features in common with the royal landscapes of Kew and

Richmond and was similarly located on the rural fringes of London. As well as the botanic collections and agricultural experimentation, both also included astronomical observatories, with scientific instrument collections, which will be explored in more detail in the next chapter.

The increasing pace of enclosure in the eighteenth century meant that landowners appropriated land for forestry, arable crops, and animal production, and therefore often took an associated interest in land-based economics. However, there was a blurring between economic necessity and demonstrations of status. Everything within the landscape park had a variety of overlapping functions, so the attractively arranged sheep on the hillside were ornamental objects as well as utilitarian grass cutters, and along with their fleeces, also eventually destined for the market. Therefore, the shared interest of the landowning classes in the breeding of livestock for improved characteristics mirrored their interest in the improvement of other less agricultural animals, such as horses and hounds.[17] As Clare Bucknell argues, this intellectual interest in economics and improvement in the mid-eighteenth century also "marked a transfer of activity from parliamentary intervention and amateur enthusiasm to the more sophisticated efforts of an emerging 'rural professional class.'"[18] Medical practitioners, with their rising social status and specialist botanic and agricultural expertise, were also engaged in this economic enterprise and formed an important group within this emerging class.[19] They were not alone, though; other professionals such as lawyers were also developing their polite standing by publishing works on agricultural improvement.[20] For example, Thomas Ruggles, a barrister, published a series of essays on "Picturesque Farming" in Young's *Annals of Agriculture* in the 1780s.

"Men well versed in the science as well as practice of agriculture"

Medical practitioners, particularly the wealthier physicians, were particularly well placed to capitalize on this growing interest due to their training. Young keenly promoted the utilitarian aspects of agricultural experimentation and strongly argued for the need for specific agricultural education as part of the university curriculum. He questioned, "But where is the botanical, chemical and mineralogical knowledge? To come more immediately to the point, where is the instruction in agriculture, and the arts immediately connected with, or dependent on it?"[21] Despite his denial, many of the building blocks necessary were already being taught within medical faculties, as students undertook courses in chemistry, botany, and comparative anatomy, as has already been discussed in chapter 1. Medical practitioners were then well placed to take what Young described as an essential scientific approach to agricultural improvement. In 1794, John Monk, in one of the county reports for Young's governmental Board of Agriculture, summed up this approach by arguing that "nothing has retarded improvements more than noblemen and gentlemen of large fortunes employing stewards who are ignorant of the principles of agriculture; they ought always to be men well versed in the science as well as practice of agriculture."[22] This also suggests a class dynamic in which the more practically trained steward was now overlooked as a custodian of crucial knowledge in favor of more learned and polite approaches to agricultural developments.

However, a direct relationship between agricultural experiment in domestic spaces and the type of specialist education medical practitioners received can be seen in this example from the Glasgow physician William Cullen. Writing to one of his counterparts in London, the physician William Hunter, in 1788, Cullen opined, "I have got a farm, and if the public would not laugh, I would call it a villa. It is truly a scheme of pleasure not profit. I hope indeed to make two stalks of corn grow where one grew before; but I believe this will be of more benefit to the public than myself and my purpose is purely the beauty of strong corns and fine grass."[23] Cullen was conducting agricultural experiments on his own farm and one belonging to his brother near Glasgow. The findings from these experiments were then translated and disseminated via his university lectures on agriculture and agricultural chemistry for students, the majority of which were studying medicine. This demonstrates both his interest in the scientific basis of agricultural production and the importance of the transmission of his research. As Charles Withers has demonstrated, this approach raises Cullen's profile to someone who was an important figure in agricultural circles during this period as well as a medical practitioner and lecturer.[24] Attention to Cullen as a key figure reveals the interlinking of interests in medical and scientific practice, agricultural improvement and national benefit.[25] This also underlines the close connections between medical practitioners, the subjects of chemistry and botany, and the production of scientific and agricultural knowledge on their domestic estates. This experimentation was seen to be of benefit to students, with such knowledge making its way back into the classroom, as well as to the wider public when findings were more widely disseminated.

There are evidently strong overlapping threads here between medical knowledge, natural history, and attempts at agricultural improvements. As Simon Chaplin notes, "Collections of natural history were commonly portrayed as being pertinent to specific areas of practical study—medical botany or mineralogy, for example—and were closely connected with more explicitly utilitarian functions, such as schemes for improvement in medicine, agriculture or mining."[26] In this way the botanic collections, museums, libraries, and landscape-based experimentation can all be seen as intertwining in order to link the practical study with the utilitarian function.

An anonymous text published in 1760 argued that in order to educate farmers, those physicians who were interested in making agricultural improvements should be set in "convenient farms almost in every district in the country . . . and called upon the physician to use the leisure spared from raising wholesome food for the preservation of health, and from cultivating herbs necessary to cure disease, in improving manures and adapting plants to proper soil."[27] This suggests there was an expectation on the part of the writer that medical practitioners should use their leisure time to study soil, because better food supplies and botanical remedies could potentially reduce the workload of physicians. As well as linking poverty and food insecurity with disease, this was also grounded in the understanding that medical practitioners had specialist knowledge in both botany and chemistry, so they were ideally placed to conduct this work.

This specialist knowledge can most directly be observed in the experimental growing of plants with medicinal uses. For example, depicted just outside the walls in the 1777 plan of John Hope's Leith Walk garden in

Edinburgh, there was an experimental field containing around three thousand rhubarb plants.[28] Rhubarb was prized for being a mild but effective remedy, and throughout the eighteenth century it was viewed as a valuable commercial product.[29] For example, the Society for the Encouragement of Arts, Manufactures and Commerce offered prizes for growing rhubarb, and it is clear from their archives that many doctors were conducting botanical and medicinal trials. In 1785 Dr. Collingwood wrote to the society outlining his table of the comparative strengths of various different rhubarbs, which he had "tried on patients of all ages in the years 1779, 80, 81, 82, 83, 84 & 85 at Norham and Alnwick."[30] He records that they had used the rhubarb for a variety of conditions, including "general debility, acidity and griping, flatulency from cold, rheumatic in stomach, colic with pain, wont of appetite, pain in head from stomach."[31] There are indications that he had been growing the rhubarb plants himself, as he notes that "the medical virtues are greatest on a light gravelly soil lying to the sun."[32]

Evidence of this type of home experimentation can also be seen in letters written to Lettsom from correspondents abroad. Dr. Benjamin Rush, a physician, social reformer, and signatory of the American Declaration of Independence, wrote to Lettsom in October 1788 stating that "Capt. Sutton will deliver to you a small bundle, containing samples of two medicines, which have lately become parts of the Materia Medica of the physicians of the United States."[33] One of these specimens he notes was a plant known as "cow-tongue" in Maryland. He wrote, "I have sent you a small quantity of its seed, which I hope to hear you have sowed in your garden at Camberwell. Should you succeed in cultivating it, I am sure your countrymen will have reason to thank you for it, for I know few more valuable medicines."[34] Grove Hill, then, offered a space for the cultivation of potentially valuable medicinal plants—in this case from the American colonies—which could then be circulated to others in a similar way to botanic specimens.[35]

Networks of gardens feature more overtly as spaces for experimentation in a letter written in 1797 by Dr. De Salis of Wing in Buckinghamshire to the Society for the Encouragement of Arts, Manufactures and Commerce. He detailed how he had dried the root of *Rheum palmatum* and that he had taken cuttings from the root which he had planted in separate pots "filled with Garden Mould."[36] He recorded that "in the Spring they will be turned out in a soil proper for them, where they are to remain," which was presumably in his own garden.[37] Other gardens were also crucial to his work. He wrote that he had applied to a number of gardens in the neighborhood of London for seed, including the Duke of Northumberland's garden at Sion. Only the latter successfully germinated, and he concluded that only very old plants produced viable seed. Gardens then formed a network in the exchange and experimental growing of medicinally commercial plants from around the globe, which paralleled their botanical and other agricultural benefits, and who better to grow and observe these plants than medical practitioners with their extensive training?

This use of the domestic garden for close observation turns what is generally seen as an ornamental space for pleasure and leisure into a scientific laboratory for experimentation and scrutiny of the natural world. John Hunter demonstrates this approach, as we shall see later in his garden at Earl's Court. But as a brief example here, he recorded his close observations of the *Mimosa pudica,* a plant that could move its leaves (fig. 5.2): "In order to have the greatest part of the day before me, I began my experiments at 8 in the morning, while the leaves were in full expansion, and I continued them till 4 in the afternoon, as longer would not have been just, for they begin to collapse of themselves between 5 and 6 o'clock."[38] This level of surveillance of a variety of plants would have been facilitated by a close proximity of the plant to the domestic residence, most likely in the garden or an adjoining greenhouse.

This use of the garden as an interim experimental space between the field and the laboratory is also alluded to in a review of Coyte's *Hortus Botanicus Gippovicensis,* in *The Gentleman's Magazine* of 1796, where the reviewer notes the addition of a discussion of the grasslands of Suffolk. They describe how "four large *plats* of *Tannington Green,* brought to the Doctor in the winter, taken as far distant from each other as the common, which contains nearly 200 acres, would properly admit of, and *planted* near his residence; that whatever plant made its appearance might be constantly under examination, and minuted down at the time of its coming."[39] They conclude that this "is at least a *new* way of botanizing."[40] The main aim of

Fig. 5.2. The moving plant, mimosa (*Mimosa pudica*), is depicted here as "sensibility," with an image of Emma Hamilton striking one of her famous attitudes demonstrating the fashionable interest in the plant. Stipple engraving by R. Earlom, 1789, after G. Romney. Wellcome Collection, CC BY 4.0.

Coyte's experiment was to analyze the quality of butter in relation to the grass used on dairy farms, and for this purpose he listed "twelve species of grass, one rush, one sedge and eleven species of broad leaf plants" that would have been found in local grassland.[41] From Coyte's letters to James Edward Smith in 1788, it is evident that as well as collecting a wide variety of plants from the wild for his collection, he also received the seeds of "*Scorzonera hispanica* (viper's grass) from Mrs Hasell, wife of an eminent

Ipswich lawyer" as well as plants from Botany Bay.[42] Blatchly and James suggest that these were most likely to have been gifts from Banks, who was frequently mentioned in his letters to Smith.[43] Plants, then, could be of interest whether locally sourced, shared between botanic networks, or sent from farther afield. The domestic garden created a space where all these productions of nature could be investigated and their significance assessed.

The interest in grasses by medical practitioners was not limited to their own private gardens. Curtis, who also published a work on grasses in 1790, designed beds within the London Botanic Garden specially designated for species of grass. Similarly, over at the Cambridge University Botanic Garden, Charles Miller, the gardener and son of Philip Miller the famous gardener at the Chelsea Physic Garden, was reported as conducting experiments on wheat in 1768. According to Dr. Watson, who communicated the results of these trials by letter to the Royal Society, these experiments were ongoing, but he noted that Miller had already achieved the raising of two thousand ears from a single grain through the division of the plants as they grew.[44]

Curtis, perhaps unsurprisingly, claimed that the importance of advanced botanic expertise in such agricultural experimentation was essential. He argued that distinguishing between types of grass was challenging, even for trained botanists, so if they were "often at a loss to know some of them apart; if so, how easily may the husbandman be deterred from the arduous task."[45] He also condemned the work of nonspecialists, stating that "grasses as well as other plants, have been frequently recommended from a partial and limited observation of them, by persons who neither knew them well as botanists or agriculturalists, or who recommended them, merely to gain by the credulity of the public."[46] Curtis, like Monk, saw himself in a position of knowledge with suitable means, including the garden space "to make the experiment," which he hoped would be an essential service for the public and prove a "great national advantage."[47] In this way the local knowledge obtained by those employed in agricultural labor is overwritten by the claim by Curtis and others to a superior kind of scientific knowledge.

This importance of the right kind of botanic knowledge is again highlighted in the preface to his *Flora Londinensis,* in which he argued that "a knowledge of the plants themselves is first necessary, and for want of which, indeed, the experimental farmer cannot effectually communicate his improvements."[48] He of course claims this essential role as botanical expert for himself. Medical practitioners such as Curtis, due to their training (in his case, through his development as an apothecary's apprentice), could claim advanced botanic expertise, and those with rural estates or access to space in botanic gardens could work with gardeners to trial and experiment on a variety of plants and animals, which they hoped would prove to be of economic and scientific value. However, the labor of those doing the horticultural and agricultural trials is often not visible in the tracts written by those directing the work, again reflecting the invisible knowledge and labor inherent in all these experimental landscapes.

Plants themselves were not the only elements of interest. Understanding and improving soil as a material that supported their growth was also a concern. The physician Francis Home, professor of materia medica at the University of Edinburgh and president of the Royal College of Physicians of Edinburgh, won a prize for his 1756 treatise *The Principles of Agriculture and Vegetation,* in which he applied chemistry to farming. In particular, he described "the growth of plants potted in soils treated with compounds like magnesium sulphate and potassium nitrate" and also "showed that plants gain nutrition from air."[49] The York physician and founder of the Agricultural Society of York, Alexander Hunter, published a work titled *Georgical Essays,* which was a central reason for his election to the Royal Society.[50] It similarly concentrated on the qualities of soil, as well as wider agricultural themes and natural history.

So prevalent was this interest in soil and compost among medical practitioners that Edward Jenner (plate 22), better known for his work on the smallpox vaccine, also conducted experiments on various types of fertilizer. On June 5, 1787, Jenner wrote to Banks describing various experiments he had been doing since 1780 on whether blood was a useful additive that would increase soil fertility.[51] According to this letter, in February, "a small quantity of the Serum of human blood was pour'd over about a square foot of grass on a grass-plot. Three sprinklings were given at the distance of a fortnight each, and the whole quantity applied was the serum contain'd in forty Ounces of blood."[52] By April, he records "that the effects it has produc'd on the vegetation of the grass is astonishing. It is beautifully green & thick & has sprung up several inches while the surrounding grass has but just begun to shoot."[53] The use of blood had a less positive outcome on polyanthus plants, as, "at about the time when the flower-stems (which were uncommonly vigorous) were push'd up to about half their height, they suddenly wither'd away & died."[54] Similarly, the peach trees that did best were those that were fertilized with animal manure. He conducted variations of the same experiment on currant trees and mustard seed.

His natural history observations also led him to consider the important role of earthworms for gardeners. Humphry Davy, the chemist and inventor, recalled a conversation he had had with Jenner in 1809 on the habits of animals. Unlike Jenner, he said, he was "more disposed to consider the dunghill and putrefaction as useful to the worm, rather than the worm as an agent important to man in the economy of nature."[55] However, Jenner viewed earthworms as essential in the creation of manure for plants. In his recollection of this conversation to his brother, Davy remembered Jenner arguing that "they act as the slug does in furnishing materials for food to the vegetable kingdom; and under the surface, they break stiff clods in pieces, and finely divide the soil. They feed likewise entirely on inorganic matter, and are rather the scavengers than the tyrants of the vegetable system."[56] This concern with improving the economy of nature, an essential feature of agricultural and horticultural science, was facilitated by observations and experimentation in domestic spaces.

The transformation of rural estates and their landscapes can only really be understood in reference to the ideology of improvement that was all-pervasive during the eighteenth century.[57] Broad ideas of late-eighteenth- and early-nineteenth-century "improvement" impacted on a range of features, from agricultural interests through civic architecture to workhouses and prisons. As Sarah Tarlow argues, "The economic and moral meanings of the term became increasingly knitted together so that by the mid-eighteenth century "Improvement" meant both profit and moral benefit."[58] Using her definition, the late-eighteenth-century garden can be viewed as a space able to bring both economic and moral benefits. They were educational, with their attempts to classify and disseminate information about plants, as well as providing spaces where economically viable plants could be showcased alongside methods for increasing crop production.

As Williamson has noted, "Contemporaries used the term 'improvement' indiscriminately for the reclamation of 'waste,' for schemes of afforestation, and for the laying out of parks and elaborate pleasure grounds."[59] Reflecting this dual interest in profit and ornamental improvements, the twin terms *dulci* and *utile* were often expressed by those writing about botany and agriculture in this period. As Curtis asserted in 1778, botany was "among all the studies which engage mankind, . . . none more pleasing, more extensive, or in which the *utile dulci* is so intimately blended."[60] This interest in using domestic gardens for both ornament and utility reflected the highly fashionable and ornamental style of farm, known as a *ferme ornée,* which emerged in the eighteenth century. One of the archetypal gardens designed in this manner was by William Shenstone at the Leasowes in Shropshire. Cullen, for example, evidently saw this as a model landscape and wrote, "I hope, in short, in a few years to shew a Leasowes in Scotland."[61] Hope also visited the Leasowes and was interested in its design, which suggests an interrelationship between botanic gardens and ornamental farms at this time.[62] Similarly, Weston in 1773 argued that for gentlemen to raise a profit from their labor, "it must be by uniting the garden-culture with farming."[63] This blurring of the farm and the garden is a key element of these experimental estates.

These attempts at unity between the garden and the farm were related to wider concerns regarding the production of natural knowledge. The interest in natural knowledge and its display more broadly was similarly about both use and beauty.[64] This approach to combining the ornamental with the productive was heavily influenced by a reinterpretation of the classical works of Virgil. For example, John Evelyn wrote to Sir Thomas Browne in 1660 to demonstrate how the ideal garden combined pleasure with moral and economic use following the Virgilian estate laid out in *Georgics,* which provided both an agricultural model as well as a justification for the high level of botanical experimentation that was already occurring.[65]

This model appears to have influenced the design and use of country estates by medical physicians, where the botanic and economic were integrated with the ornamental and pleasurable. As with other more elite "improvers" of the landscape garden, the Virgilian concerns of "agricultural improvement, botanical experimentation, philosophic speculation, rural retirement, and arcadian landscape" could be found in the more modest estates of men such as Lettsom.[66] The direct influence of Virgil on these eighteenth-century medico-botanists can be seen in the case of John Martyn, the professor of botany at Cambridge, who in 1742 published *Flora Virgiliana,* which was a translation of the first two books of the *Georgics.*[67] Similarly, Alexander Hunter's work *Georgical Essays* also reflected this classical influence.[68]

This interest in agricultural experimentation with new plants and improved methods of cultivation was also encouraged by the founding of societies with economic interests in the eighteenth century. These included the Society for the Encouragement of Arts, Manufactures and Commerce, established in 1754, whose motto was *Utile et Dulce,* and the creation of regional agricultural societies, including the Royal Bath and West of England Society, in 1777, which had its own experimental garden. As with the central collection and dissemination of the mangel-wurzel seed discussed earlier, these societies facilitated the circulation of knowledge regarding agricultural developments as well as actual botanic material. For example, Reverend Dr. Thomas Lyster wrote to the Society for the Encouragement of Arts, Manufactures and Commerce thanking them for hempseed on March 13, 1786.[69] The London-based society had sent the Dublin Society "two quarts of China Hemp Seed for Experiments," and Lyster stated that "when Experiments are made with it, our Society will with pleasure acquaint yours with the Benefit."[70] In this way seeds could be shared across the botanic network, between societies and institutions as well as between individuals, and knowledge would be fed back to the center.

The Dublin botanic garden had been established by 1795, so it is possible that the hemp was grown there rather than in a domestic garden, although this is unclear. The focus of the new botanic garden at Glasnevin was certainly agricultural, partly due to the involvement of the Dublin Society, established in 1731 to improve the poor economic condition of the country by promoting agriculture, arts, industry, and science in Ireland.[71] As we have already seen, the foundation of this garden had nationalist overtones that were highlighted through the rhetoric of "improvement." This reflected some of the impetus seen behind the calls for new botanic gardens in developing British urban centers such as London and Norwich. However, like all designed landscapes, Dublin's garden also reflected local concerns specific to the region and the players involved.

"Trying many experiments": John Hunter at Earl's Court

The use of the garden as a specialist space for experimentation and observation is best demonstrated by a consideration of the rural estate of surgeon and anatomist John Hunter (fig. 5.3) at Earl's Court. Like Lettsom and others, Hunter had a central London building, in Leicester Square, which housed his domestic quarters, anatomy school, and an anatomical museum collection. From here he also ran his surgical practice, but in common with many of the other practitioners in this book, he also bought two acres of land in the rural village of Earl's Court on the outskirts of London in 1764 (plate 23).

In 1793 Thomas Baird visited the Earl's Court estate as part of his research for his *General View of the Agriculture of the County of Middlesex.* He was employed on this venture by the Board of Agriculture and Internal Improvement. In a section headed "Important Experiments," Baird described Hunter's estate as "the villa of John Hunter, the celebrated surgeon, who is trying many experiments, which may be of considerable service, both to the gardener and the husbandman."[72] Here Hunter was singled out for attention due to the wide use of his garden for horticultural and agricultural experimental activities. The Baird report is particularly important as it was reprinted widely in the London newspapers and thereby a description of both the estate and Hunter's activities was disseminated to a broad audience.[73] There is, however, no evidence to suggest that Hunter opened his gardens to visitors, so this was presumably for scientific interest rather than garden publicity.[74] This is not to say that his experimental work was invisible to the public, as many of the resultant preparations of anatomical

Fig. 5.3. Photograph of John Hunter's 1786 portrait by Sir Joshua Reynolds. This depicts Hunter with some of the numerous anatomical specimens he collected, which now form the basis of the Hunterian Museum at the Royal College of Surgeons, as well as a group of anatomy textbooks. Wellcome Collection, CC BY 4.0.

material made at Earl's Court ended up displayed in his Leicester Square museum.[75]

Hunter's domestic research had clear connections with his wider interest in scientific experimentation, which included, among many other things, the successful transplantation of a human tooth into a cockerel's comb, as well as a range of other human and animal-based experiments on the generation of body heat, venereal disease, and the transplantation of other organs.[76] It is perhaps then unsurprising that he took the same experimental approach to the plants within his garden at Earl's Court. Baird's report stated that Hunter was "very curious in plants and has in his green-houses and hot-houses a great variety of the most choice and rare productions of nature, in the collection of which he has neither spared pains or expense."[77] He went on to record that Hunter was experimenting with forest trees in order that "he shall be able to direct or determine the growth of trees . . . to any particular part of the trunk he may choose. For example, if from an oak, a plank is wanted of a given length and of an equal breadth at both ends . . . he is of the opinion that the tree may be trained and disposed to grow in such a matter that it will yield the plank of the exact dimensions required."[78] When the demolition of the house occurred in 1886, remarks were made by observers of the material features still visible from his experiments with grafting different tree species together, such as "a rough-skinned oak, with smooth-skinned branches grafted on to it."[79] This interest in trees corresponded to a general interest in agricultural improvements and the use of landscape features such as forests for economic purposes as well as aesthetic appreciation. For example, Alexander Hunter, no obvious relation to John, as well as publishing the *Georgics,* also published a revised version of John Evelyn's *Silva* in 1776.[80]

John Hunter was also conducting trials regarding different types of growing material for plants in line with the types of experiments conducted by Home and Jenner among others, as discussed earlier. As noted in chapter 2, Hunter also conducted experiments on animals on his estate, and many of the creatures listed by Stephen Paget in 1897 were domestic animals, again suggesting an agricultural focus. It would also seem that Hunter's interest in the production of heat by animals and vegetables was an important factor in the living material obtained and the subsequent experiments conducted.[81] Hunter wrote that he conducted some of these experiments to "ascertain whether vegetables could be frozen, and afterwards retain all their properties when thawed, or had the same power of generating heat with animals."[82] Baird certainly had more of a pastoral slant on the scenes that he witnessed in 1793, although as Hunter was sixty-five and very successful, this may represent a point in Hunter's life when he had an increased amount of time and money to spend on developing the estate.

Baird recorded that "the variety of birds and beasts to be met with at Earl's Court . . . is a matter of great entertainment. In the same ground you are suprized [*sic*] to find so many living animals, in one herd, from the most opposite parts of the habitable globe. Buffaloes, rams and sheep from Turkey, and a shawl goat from the East Indies, are among the most remarkable of those that meet the eye."[83] Although Baird stated that they were a matter of entertainment, thereby implying the animals were an element of spectacle within the landscape, the most exotic that he described on this visit were still animals that were bred first and foremost for wool and meat production. These domestic beasts are the most prevalent in Baird's description, and the vaults built around the house, described by Buckland on his 1875 visit to the estate, were depicted as follows: "Mr Hunter built his stables half underground; and that he also had vaults in which he keeps his cows, buffaloes and hogs."[84] It is evident from these portrayals that much of Hunter's interest in these animals related to crossbreeding for agricultural purposes. Although Baird necessarily focused on the agricultural aspects of Hunter's work, this does suggest an agricultural estate rather more than a menagerie, at least in the 1790s.

One material piece of evidence that relates to a more exotic tale of animal husbandry is the mound that was located in the garden to the back of the house. Whether or not the mound was an artefact of a previous age or built by Hunter himself, we know that he utilized it and that it remained in the garden until the demolition of the house in the 1880s. There is a watercolor in the Hunter family album (plate 24) which suggests that it had a pastoral feel and that during Hunter's time the mound was used as an animal pen, even if it was not originally designed with that use in mind. As noted earlier, this type of mixed use fits the style of eighteenth-century landscape design known as the *ferme ornée,* or "ornamented farm," where utilitarian buildings, such as cowsheds, could also be attractive features acting as eye-catchers or decorative structures within the landscape.[85] So Hunter's use of a physical feature for aesthetic and pastoral purposes was in keeping with the aesthetic taste of the period. The Greater London Council survey published in the 1980s described it as a "mound containing vaulted byres for the larger animals."[86] The mound was referred to as the "lion's den" by Buckland, again in 1875, and although there is no record of Hunter housing a lion, there is a narrative regarding Hunter and the keeping of leopards at Earl's Court. As Wendy Moore relates, "The leopards once broke free from their chains and ran into the yard where they attacked the dogs. . . . Somehow he managed to catch the animals and get them back in the den."[87] However, whether the den is in fact the mound or some other feature is unclear, and it has been suggested that the mound was actually used to keep Hunter's buffaloes, which, although exotic, were also used as draft animals.[88]

As well as the agricultural investigations recorded by Baird, Hunter also conducted other botanical and natural history experiments, as recorded within his letters and published articles. Notable examples that were conducted at Earl's Court include an attempt to culture pearls in the garden pond, and his research on bees.[89] His observations on bees perhaps best demonstrate how domestic gardens could be used for close examination and experimentation with the natural world.

Promoting a Bee Society

This use of close observation in a domestic space is best demonstrated by considering the beehives constructed within the conservatory adjoining the house at Earl's Court (fig. 5.4), which enabled Hunter to conduct detailed observational work on the habits and behaviors of bees. This in turn led to the publication of a scientific paper, "Observations on Bees," which was the last work he contributed to *Philosophical Transactions,* the journal published by the Royal Society, in 1792.

In order to conduct detailed observations, the hives were constructed to his own specifications so that they had "different panes of glass, each pane opening with hinges so that if I saw anything going on that I wished to examine more minutely or immediately, I opened the pane at this part and executed what I wished, as much as was in my power."[90] He recorded his observations as follows: "When I saw some operations going on the dates or periods of which I wished to ascertain, such as the time of laying eggs, of hatching, &c. I made a little dot with white paint opposite to the cell where the egg was laid and put down the date."[91]

Fig. 5.4. Photograph of Earl's Court House taken just before its demolition in 1875, showing, to the right of the house, the conservatory in which Hunter kept his bees. This image, like plate 25, is pasted into the Hunter family album. From the Archives of the Royal College of Surgeons of England.

His close observational work is particularly well demonstrated by his detailed attention to how bees produce their distinctive buzzing sound:

> Bees may be said to have a voice. . . . But they produce a noise independent of their wings; for if a bee is smeared all over with honey, so as to make the wings stick together it will be found to make a noise, which is shrill and peevish. To ascertain this further, I held a bee by the legs, with a pair of pincers; and observed it then made the peevish noise, although the wings were perfectly still: I then cut the wings off, and found it made the

> same noise. I examined it in water, but it then did not produce the noise, till it was very much teased and then it made the same kind of noise; and I could observe the water, or rather the surface of contact of the water with the air at the mouth of an air-hole at the root of the wing vibrating. . . . I have observed that they, or some of them, make a noise the evenings before they swarm, which is a kind of ring, or sound of a small trumpet: by comparing it with the notes of the piano forte, it seemed to be the same with the lower A of the treble.[92]

This detailed multisensory analysis that included the use of vision and sound gives a sense of the ways in which Hunter used his domestic space as a laboratory. As a man who once described his head as "like a beehive," he was clearly fascinated by all aspects of the natural world. This deep interest in bees was no doubt also related to his agricultural interest in animals, compost, silkworms, and trees—they were all important in relation to the economic value of an estate. This was of course not new. For example, in 1651 Richard Childe described that deficiencies in beekeeping meant that the potential profits that could be made from products such as honey and wax were not being reached.[93] In the eighteenth century more productive beekeeping was encouraged by the Society for the Encouragement of Arts, Manufactures and Commerce, which awarded agricultural premiums (financial prizes), including forty-one for "bee hives, collection of wax, or more effective management of bees."[94]

Hunter was clearly then not alone in thinking that bees were important objects of study. In a somewhat different and perhaps more agrarian approach, Lettsom also kept bees at Grove Hill. His apiary (fig. 5.5), unlike Hunter's scientific research station, consisted of "sixty-four hives, each of which was distinguished by the name of some kingdom, or independent nation, commencing with the North of Europe, afterwards including Asia, Africa and America, and concluding with the great European islands."[95] This organization of the world via beehives perhaps reflected earlier ideas of the natural order of a nation being symbolized through the arrangement of bees. Evelyn, for example, described how "they have a Citty [*sic*], King,

Fig. 5.5. Depiction of some of the hives in Lettsom's apiary at Grove Hill. The image also includes Saint Paul's Cathedral in the background, drawing a visual connection between morality and the behavior of bees. From Maurice's *Grove-Hill.* Wellcome Collection, photo by author.

Empire, Society" and that of all insects he considered them to be "the most affected to Monarchy, & the most Loyall, reading a Lecture of obedience to Rebells in every mans Garden."[96] At Grove Hill, the various descriptions suggest that they were less representative of national or state organization and rather an illustration of global trade, or at least the trade of the empire, through a series of hives representing economic activity. These concepts are emphasised by Maurice, where he describes Lettsom's apiaries as "One mighty Empire, one pervading mind. / No civil discords in that Empire rage."[97] He also states that through Lettsom's apiary, "a kind of history of the world is exhibited in the habitations of the industrious bee," and the illustration used to accompany this element of the poem includes Saint Paul's Cathedral, possibly as a moral representation of the City of London.[98] However, overall the representation of trade chimes with the earlier references to the mercantile nature of the city with the view of ships on the Thames, and thus within Grove Hill there are clear connections to economic prosperity and commerce.

Certainly, the utilitarian aspect appears to be Lettsom's greatest concern in his pamphlet *Hints for Promoting a Bee Society.* In this tract he bemoans the establishment of societies for the improvement of the beauty of the pigeon as well as those for "fancy birds, flowers, and other trivial objects."[99] He sees the use of bees predominately in terms of economic advantage, rather than an allegory for governance, and is concerned with the decline of beekeeping in urban areas. He argues that in growing urban centers bees were "left without due patronage; and, from neglect, the stocks are annually diminishing" and suggests that premiums be given (as they eventually were by the Society for the Encouragement of Arts, Manufactures and Commerce) for research that would find the "food most suitable to bees, the best mode of taking honey, constructing hives and preserving its denizens."[100] His note that bee stocks were annually diminishing also suggests a long history of the decline of bee populations, at least in metropolitan areas. Beekeeping could, in Lettsom's view, provide "a little honey on bread," which "would save the use of butter on the occasion and be more wholesome: it is at the same time a luxury, that every family, in possession of a garden, may command without expence, and certainly with the addition of rational amusement."[101] Here we have both an economic argument and a moralistic tone, with the idea of rational amusement heralding the later discussions for the role of public parks in the nineteenth century as morally improving places of rational recreation.

Such moral lessons are woven between a citing of the economic benefits throughout the pamphlet (fig. 5.6). He describes how "where ornament and pleasure have been particularly studied, neat mahogany and glass hives have been constructed in the windows of dwelling houses."[102] He goes on to describe how this "means company in a sitting room may see into the glass hive, and be amused by the activity and labour of the industrious community every moment of the day, and learn a lesson of employing their own moments to the most useful purposes."[103] This has clear connections with the minister William Mewe's seventeenth-century transparent hive, which would similarly allow better management of the bee population as well as offering an example of sound moral and political organization. As Raylor notes, in the early modern period bees were generally "regarded as sound

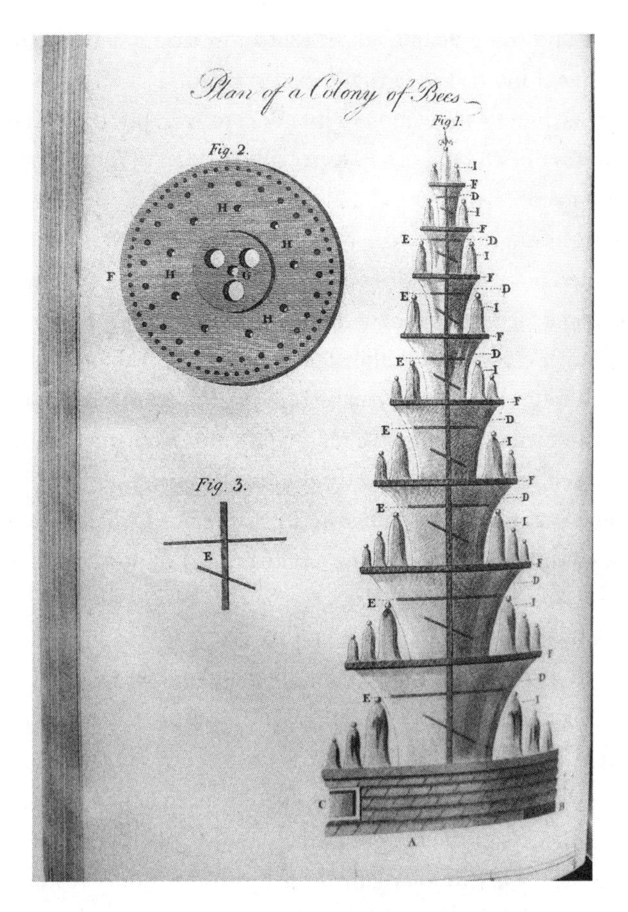

Fig. 5.6. Lettsom's plan for the ideal beehive construction, from *Hints for Promoting a Bee Society* (London: Darton and Harvey, 1796). Wellcome Collection, photo by author.

economists and good husbandmen. They were of impeccable ethical character, being clean, chaste, pious and industrious."[104] These are all elements that surely appealed to Lettsom as a Quaker, with his interest in agricultural economy and improvement in all its forms. Maurice appears to reflect this when he writes in the poem of Lettsom's bees, "Their vigorous industry, their loyal zeal, / Their generous ardour for the public weal."[105]

Beekeeping in the early modern period, then, "seemed to offer a perfect remedy for the related problems of unemployment, poverty, wealth distribution and trade deficit."[106] In Lettsom's case, his beekeeping along with his general interest in agricultural production in all its forms can be considered as directly in relation to his concern with poverty and health. As Hunting notes of Lettsom, "Diet, especially the diet of the poor, was one of his chief concerns, working as he did among the sick poor of London during years of shortages due to the Napoleonic war with France."[107] His concern with diet, poverty, and disease led to his publication of pamphlets outlining his plans for soup kitchens as well as solutions for reducing the high cost of bread and recipes for affordable sustenance, including his own for Camberwell soup. His garden can be read as another means, along with his pamphlets and letter writing, of encouraging improvements in a wide range of areas, which in turn he hoped would lead to a healthier nation overall.

This was a practical response to the problems he had witnessed first-hand when visiting the poor in the East End of London. In his 1773 pamphlet *Of the Improvement of Medicine in London, on the Basis of Public Good,* which was so popular it was reprinted in 1775, he argued that "the poor, from the occasional want of employment and wholesome food, from exposure to all changes of the weather, and from various other causes, are often visited with sickness, as well as with poverty; one, indeed, is consequent upon the other."[108] In this pamphlet he directly tied the health of the poor to the success of the nation. By helping the poor, he argued, "health, which is so necessary to their subsistence, will be sooner restored, famine and a prison avoided, the nation inriched by industry, and a hardy race of useful members preserved to the community."[109] In this way Lettsom's garden can be seen as both a scientific and agricultural space, as well as reflecting his broader concerns considering the health and status of the nation as a whole.

SIX

This "Terrestrial Elysium"

Sociability and the Garden

IN MAY 1804, TWO MONTHS before Lettsom's daughter's wedding, Grove Hill formed the backdrop for a lavish and sociable entertainment for more than eight hundred invited guests. Beginning in the house, a suite of seven rooms were thrown open to the invited throng. An anonymously penned report detailing the party was published in *The Gentleman's Magazine.* This described how there was an absence of "music, singing or cards" but rather the guests were invited to take "rational pleasure" in each other's company and Lettsom's collections for the evening.[1] This presumably was aligned to Lettsom's Quaker beliefs, as during the eighteenth century, Quakerism as a movement became increasingly concerned with plainness and encouraged the avoidance of worldly pleasures.[2] Quaker feelings aside, though, it also allowed him to encourage a scholarly appreciation of his museum, library, and garden among his guests. We are told that these included many professional men involved in "law, physic and divinity," as well as women of "genuine beauty and unaffected elegance of dress."[3]

On arrival the guests were first invited to peruse the museum and library, before being led into a specially constructed room in the garden that was seventy-two feet long and thirty feet wide.[4] There the guests were shown to tables that were

> filled with every thing desirable to the sight or the palate—strawberries still growing on the living plants—iced creams of every sort and flavor—rich jellies—confectionary of the most ingenious devices, many of the articles inclosing well-adapted

> posies;—with the more substantial dishes of ham, veal, beef, &c., &c.;—in short, such an abundance of every delicacy, as left nothing either to be wished or desired.[5]

Here the boundaries between the interior and the garden were blurred, and the productions of the gardens formed the centerpieces as well as the background for an elaborate feast to satisfy all the senses.

Reflecting the garden itself, the party assumed both a rustic and an exotic flavor through the interior decorations used in the garden pavilion. The magazine account described the ceiling as being obscured by boughs of trees and shrubs that had been freshly brought in, along with the strawberry plant centerpieces, all presumably from the immediate garden and wider Camberwell landscape.[6] To excite the guests and encourage a sense of the exotic there were fully grown orange trees, placed so that they appeared to support the roof over the diners, which would have created a fashionable pastoral setting for the banquet set before them. This use of natural elements to create theatrical backdrops had clear parallels with other more lavish events held by wealthy landowners earlier in the eighteenth century, which again accentuates Lettsom's polite status as a fashionable gentleman and his engagement with the sociable activities of the day.[7]

As always Mrs. Delany is an excellent source for descriptions of the entertainments of those moving in elite circles. In 1752 she wrote to Mrs. Dewes describing a lavish scene being erected for the Duke and Duchess of Dorset at the playhouse, presumably in Dublin. She records how the great "play-house" had been converted into a ballroom and exclaims that this "*room represents a wood.*"[8] She goes on to assess the rustic creation, stating that

> the right hand, from the portico to the end of the stage is diversified by rocks, trees and caves, very well represented. On the left hand is a jessamine bower, a Gothic temple, (which is to be the side-board,) trees interspersed, the whole terminates with a grotto extremely well exprest; three rustic arches, set off with ivy, moss, icicles, and all the rocky appurtenances; the mu-

> sicians to be placed in this grotto dressed like shepherds and shepherdesses.[9]

This bowered and rusticated space, based on familiar features of the eighteenth-century landscape garden, was designed to be used as the backdrop for a concert, a ball, and a supper. Although Mrs. Delany somewhat disparagingly notes that "the trees are *real trees* with *artificial leaves,* but when all is done it will be too much crowded to be agreeable, and most dangerous if a spark of a candle should fall on any of the scenery, which is all *painted paper*!"[10]

Potential conflagrations notwithstanding, such theatrical recreations of the landscape garden indoors were a common feature of the social whirl. A few weeks before, Mrs. Delany had witnessed a ball where as in Dublin there were singers and musicians dressed in the style of arcadian shepherds and shepherdesses. There were also ingenious methods of incorporating the theatrical garden features so that they were also a key element of the gastronomic experience. She records that "if tea, coffee or chocolate were wanting, you held your cup to a leaf of a tree, and it was filled; and whatever you wanted to eat or drink, was immediately found on a rock, or on a branch, or in the hollow of a tree."[11] Here the sensory nature of the artificial landscape was enhanced via the inclusion of exotic plant-based edible products such as tea, coffee, and chocolate. This again underlines the centrality of the imperial plant trade to eighteenth-century sociability. The design also utilized concepts of English rusticity to act as a foil to what was really a global sensation of food and drink created by the fruits of empire. These entertainments were perhaps reflective of the country as a whole, which saw itself as fundamentally rural while transforming into a more urban, modern, globally connected nation.

Another pastoral and celebratory matrimonial extravaganza, inspired by the French concept of the *fête champêtre* (outdoor garden party) (plate 25), was held at Lord Stanley's villa, The Oaks, Epsom, in 1774 to mark his marriage to Lady Elizabeth Hamilton. Mrs. Delany portrayed this as a fairy scene where guests were entertained in the garden, initially by a dialogue conducted between the usual costumed shepherd and a shepherdess, and then singing and dancing performed by sixteen men and sixteen women from the opera. Afterward the party "was employed in *swinging, jumping,* shooting with bows and arrows, and various country sports."[12] Later came the company who were dressed in costume with "the very young as peasants; the next as Polonise; the matrons dominos; the men principally dominos and many gardiners, as in the Opera dances."[13] They were then taken to a "magnificent saloon" built on the other side of the garden for supper and further entertainment, including a druid and dancing dryads.[14] These performances, both for and by the guests, combined a sanitized and polished version of rural life with a classical arcadian vision. In many ways this was in microcosm a reflection of the cultural underpinnings of the landscape gardens of the period, with their combination of classical allusion, nods to antiquarian ideas concerning hermits and druids, and idealized scenes of peaceful rurality. All of it formed a romantic backdrop for the creation of new wealth via modern trade routes and the enjoyment of the fruits of others' labor.

These artificial scenes were, however, not always created in extensive rural landscapes. In 1774 David Garrick and his wife celebrated their twenty-fifth wedding anniversary at their Hampton villa (plate 26), which was located on the fringes of London. They held an appropriately theatrical *fête champêtre* during which their garden, described by one commentator as "singularly beautiful," was illuminated with a phenomenal six thousand lamps.[15] Like Lettsom's Grove Hill, their garden was limited at six acres in size, but it could still act as a backdrop to festivities in the same manner as larger landscape gardens, and its position with easy access to the city made it ideal for those guests based in the metropolis to attend.

Despite occurring thirty years after the height of such garden entertainments, Lettsom's own 1804 dinner reflected many of these tropes but without the dancing, music, dressing up in costume, and any entertainment beyond a more scholarly and rational viewing of his collections. However, this use of the garden as a backdrop for polite events was certainly the height of sociability, and Lettsom, like Garrick, was clearly renewing and reaffirming his networks of equivalent professional men and their companions through this pastoral diversion. As Michael Brown's *Performing Medicine,* which documents the creation of the medical profession in relation to wider concerns, attests, "politeness, sociability, affability, benevolence and liberality: these were the values which underpinned the late eighteenth-century culture of medico-gentility."[16] Lettsom's presentation of his collections for investigation as well as his hospitality was a clear demonstration of his position as he exhibited his expert polite knowledge within a suitably fashionable arcadian setting (plate 27).

The evening soiree was not the only time Lettsom's garden was used for pleasurable and sociable activities. Grove Hill was also for a time the meeting place of the Athletae Club.[17] In Bransby Blake Cooper's biography of his uncle, the eminent surgeon and anatomist Astley Cooper, the author records the various clubs that Cooper belonged to and lists the Athletae as the one he most frequently attended. The club, described by Blake Cooper as "consisting of twelve professional Gentlemen, who met monthly," predominately but not solely consisting of members of the medical profession, and which was established for "the express object of recreation, and promotion of health . . . by means of active exercise."[18] This being a suitably sober version of the sociable eighteenth-century clubs, such as the Society of Dilettanti, where wine was a central element to be consumed alongside a shared, professed interest in art.

The more athletic and sober nature of the Athletae is vividly recalled in this letter from one Mr. Horatio Smith writing to Blake Cooper:

> Their post-prandial meetings were restricted to the summer months, and the earliness of the prevalent dinner hour allowed them to assemble at six o'clock, when, after pursuing their pastime till dusk, they took their tea in an alcove of the bowling-green, and separated before it was dark. At the period in question, several of the professional members retained their gold-headed canes, nor were pig-tailed wigs and cocked hats altogether discontinued.[19]

As this description suggests, the club initially met at a bowling green nearer to the center of London, where they played at quoits and bowls. Apparently when incited by Dr. Babington, they would also "occasionally engage in contests among themselves in leaping, racing, and other exercises; the

Fig. 6.1. William Hogarth's 1736 satire *The Company of Undertakers,* which depicts the medical profession with their gold-headed canes as a symbol of their status. Wellcome Collection, CC BY 4.0.

amusements of the party concluding by an unexpensive dinner at the house attached to the place of meeting."[20] Here their expert status was preserved and reinforced via sociable and playful exercises, while their professional dress was retained. In particular the gold-headed cane was an important symbol of their eminence as physicians; a cane topped with gold, silver, or ivory was the eighteenth-century version of the stethoscope and was read as an equivalent symbol of knowledge and expertise (fig. 6.1).[21] However, it would seem that as the members rose in their profession, it became seen as undignified to meet at a public bowling green, and they moved their meeting place to the more private Grove Hill, where they presumably would not be subject to the public gaze.[22] The garden was then also a space that allowed sociable exercise to take place in private.

The membership of the club was not exclusive to medical practitioners. Thomas Maurice, who wrote the laudatory poem describing Grove Hill, was occasionally allowed to join, which denotes the close and interlinking nature of these networks. Other regular visitors from outside the medical profession included John Nichols, printer, antiquarian, and editor of *The Gentleman's Magazine;* and James Boswell, friend, diarist, and biographer. In his recollections of Lettsom, Nichols remembered sociable evenings spent at Lettsom's "terrestrial Elysium," where, according to him, "good humour and sociability were the order of the day."[23] It is quite possible that Nichols was the author of the anonymous but highly detailed and complimentary description of Lettsom's 1804 party. This sociability is also reflected in Boswell's "Ode to Charles Dilly," dedicated to a prominent publisher, where he recorded that "on Saturday at bowls we play / At Camberwell with Coakley" and that "From him of good—talk, liquors, food— / His guests will always get some."[24] As with the garden party, the food and drink flowed despite Lettsom's own adherence to a more temperate Quaker lifestyle. As Nichols noted, "The good Doctor, always frugal and temperate in his personal habits, not unfrequently, after having tired three sets of horses in visiting his patients, dined at Grove Hill, and walked back in the evening to Sambrook Court."[25] Again, Lettsom's own relationship to Grove Hill and his wife was at a distance, or rather a long walk from the base of his medical practice in the City of London.

Alongside the descriptions of Grove Hill as a sociable space, there was clearly a sense that it also represented Lettsom's other virtues as a polite and learned man. Boswell uses his ode to highlight these qualities:

In Fossils he is deep, we see,
Nor knows Beasts, Fishes, Birds ill!
With Plants not few, some from Pellew,
And wondrous *Mangel-Wurzel!*
West Indian bred, warm heart, cool head,

The City's first physician:
By schemes *humane,* Want, Sickness, Pain,
To aid is his ambition.
From terrace high he feasts his eye,
When practice grants a furlough;
And, while it roves o'er Dulwich groves,
Looks down—ev'n upon Thurlow![26]

This description places the hospitality found at Grove Hill in direct dialogue with Lettsom's role as a knowledgeable man and benevolent physician. There are the expected mentions of mangel-wurzel and exotic plants as well as an emphasis on Lettsom's status, as he is again featured situated on high, surveying the world, from his seat at Grove Hill. Although clearly a place for sociable and lively gatherings such as these, the spaces were also used for smaller meetings of like-minded men from within Lettsom's professional network. Nichols stated that to his "Medical Brethren, the House, the Museum, and the Bowling-green, were always open on a Saturday."[27] Again, this suggests that the experience for estate visitors could be differentiated sometimes by class, or in this case, professional identity.[28] It also again emphasizes the semipublic nature of the collections, whether living or not, and their use by a range of audiences, not just the owner and their close family, friends, and colleagues.

Similar facets of Lettsom's identity are perhaps represented in the iconographic choice of deity to preside over the lavish May garden party. Our anonymous writer describes the party as being presided over by a statue of Minerva holding a banner on which verses were written to commend friendship (presumably both as members of polite company, as well as a reference to Quakers as Friends). This also announced that the owner was using "Nature's gifts of various kind / To gratify the enquiring mind" rather than the "cards or drum."[29] Again, this emphasizes the rational nature of the entertainment and the morally uplifting role of nature and thereby God, as opposed to the Quaker perception of the more frivolous and immoral use of music and gambling. Although it should be noted that despite Lettsom's own aversion to alcohol, the address printed within the

Fig. 6.2. Richard Wilson's black-and-white chalk sketch of the Temple of Minerva, Rome, 1754. Yale Center for British Art, Paul Mellon Collection.

description of the entertainment also noted that guests should not "spare our Cakes, our Wine, or Fruit," highlighting the luxurious nature of the event and Lettsom's adherence to some expected social codes.[30]

It is worth considering the role of Minerva as a presiding deity in more detail, as such a symbolic choice is unlikely to have been made lightly given the audience and the emblematic features to be found in the garden. Generally considered the goddess of wisdom, Minerva also has associations with medicine, government, and war. In the eighteenth century, the Temple of Minerva Medica in Rome, then understood to be a temple dedicated to Minerva the doctor, was one of the most commonly painted antique ruins. As a key example, in the 1750s it was sketched by Richard Wilson (fig. 6.2), a pioneer of landscape painting, for William Legge, 2nd Earl of Dartmouth, as part of a souvenir set of drawings recording the antiquities and historic sites the earl visited as a young man, highlighting its significance as a key monument to be visited on a gentleman's grand tour.[31]

As a fellow of the Society of Antiquaries, although without the personal experience of undertaking a grand tour himself, Lettsom would have been aware of its existence. Similarly, he would have surely known about the 1790 excitement of the excavations in Bath, which revealed a Romano British link to Minerva. These finds, which included a gorgon's head and sections of the Temple of Minerva, set the antiquarian world alight.[32] The Society of Antiquaries hosted Sir Henry Engelfield and Thomas Pownall's presentation of their findings from this excavation, and these were also published, with Englefield's paper being reprinted as an appendix to Richard Warner's *History of Bath* in 1801.[33] Lettsom may well have been using the statue of Minerva to signal his antiquarian knowledge and gentlemanly status, as well as reinforcing his status as a medical man.

Much like the development of expert knowledge in botany and other forms of natural history exhibited by medico-gentlemen, their expertise in antiquarianism could also be valuable for network building and maintenance, as the newly professionalizing class attempted to strengthen ties and gain elite clients. This was a trait shared by other professional groups, such as lawyers, who also capitalized on their networks of wealthy clients to materially build their own collections as well as cement key connections.[34] Antiquarian interests, like botany, agriculture, and natural history, all aped courtly activities and helped to raise the social standing of professional men.

As Rosemary Sweet has demonstrated, the policing of polite knowledge as an area solely for gentlemen was enforced throughout the period:

> Pretensions to antiquarian learning from those below the rank of gentleman or man of property were frequently derided, for to accept their worth would have been to open up the possibility that antiquarianism was not inseparable from gentlemanly status; gentlemen would therefore no longer be able to regard their antiquarian interests as self-evident proof of their own gentility.[35]

Medico-gentlemen were then signaling proof of their social status through displays of antiquarian and other forms of polite knowledge. There are clear parallels with the idea of advanced expert botanic knowledge being privileged over lay knowledge, as described in the previous chapter. Lettsom by opening his museum and library to his guests, as well as providing a luxurious meal and demonstrating his sociability, was also reinforcing his status as a learned gentleman.

This is further accentuated when we consider that Lettsom's domestic residence, with its agricultural experiments, botanic collections, greenhouses, museum, and library, can best be understood as a scaled-down version of Kew gardens as developed under George III. The symbolic use of Minerva may also have connected Lettsom to Kew. In the 1750s, along with the more famous Chinese Pagoda, Alhambra, and Moorish Mosque, William Chambers built a Gallery of Antiques containing a statue of Minerva. This gallery, which was open to the sky, contained ten statues representing the mythic encounter of the muses with Minerva on Mount Helicon.[36] As if to confirm Grove Hill's status as a similar, if more domestic, research institution than Kew, Lettsom also built an astronomical observatory within his garden, which was part picturesque object, part scientific research tool (fig. 6.3). The observatory can also be seen as both a courtly and scientific entity, and the whole house and garden can be read within a longer scientific history, where the botanic garden, the library, the museum, and the observatory have, often literally, resided alongside each other.[37]

Modeled on a cork miniature of the Temple of Vesta at Tivoli (fig. 6.4) near Rome, Lettsom's observatory was known as the Temple of the Sibyls.[38] Garden observatories do not seem to have been popular in this period, but we can find a precedent at Kew. The King's Observatory, designed by William Chambers, was built in 1768–1769, in order that George III could witness the transit of Venus, an activity encouraged by Dr. Stephen Demainbray, as tutor to the royal family (plate 28).[39] The designs, however, were very different, with Lettsom's being supported "by the trunks of eighteen oak trees which retained their bark and cropped branches, entwined with ivy, virgin's bower, honeysuckle, and other climbing plants."[40] This was a more rural and picturesque ornament for the landscape than Chambers's plainer Anglo-Palladian villa, and one that fully combined the fashions of the day—the scientific, the classical, and the rustic (plate 29). In terms of

Fig. 6.3. Engraving of Lettsom's rustic observatory based on the Temple of Vesta, Rome. From Maurice's *Grove-Hill.* Wellcome Collection, photo by author.

visual style, Lettsom's observatory had more in common with Jenner's rustic garden hut, which he converted into his Temple of Vaccinia, described in the next section, which was similarly functional and fashionable.

Within the observatory a collection of cork models created by Richard Du Bourg were kept (plate 30, for an example of his work), including the aforementioned replica of the Temple of Vesta at Tivoli. Other recreations in cork of various ancient features included the Arch of Titus, the Tomb of

Fig. 6.4. Undated engraving of the Temple of Vesta near Rome, which was the model for Lettsom's observatory (see fig. 6.3). Wellcome Collection, CC BY 4.0.

the Scipios, the Baths of Caracalla, Virgil's Tomb near Naples, the Temple of Health, and Stonehenge, again reflecting the themes seen illustrated at the party and within the garden itself. There were also models of medical, ancient British as well as classical, antiquaries and a link back to Virgil and his rural Arcadia.[41] This model collection was complemented by the purchase of the scientific apparatus used by James Ferguson for his public astronomical demonstrations. These scientific tools were used by Lettsom for the private instruction of his family, particularly his children, as well as used by friends.[42] Although they were purchased on Ferguson's death by the physician William Buchan and then later passed to Lettsom, the apparatus would also signify a professional and personal association, as Ferguson had been one of the signatories of Lettsom's Royal Society Fellowship application in 1773, along with Solander, Benjamin Franklin, and others.[43] Like the plants removed to Grove Hill from West Ham following Fothergill's death, the material collections were also physical manifestations and memorials of friends and professional colleagues. The botanic experimentation, the observatory with its models and scientific tools, and the collections within the library and museum confirmed Grove Hill as a sociable scientific space.

Minerva's symbolic role was multilayered and may also have been designed to stress this facet of Lettsom's character, as a leader of scientific medical practice. He certainly appears to have viewed the goddess as a definitive symbol of the developing scientific and medical practice of the time. In *The European Magazine* of 1802, Lettsom described a letter from America that outlined the new use of vaccination for smallpox among the Native American population and praised the role played by Thomas Jefferson in his presidential capacity.[44] Within the article Lettsom delineated a medal commemorating Jefferson, which he had been sent with the letter. This had the president's head imprinted on one side and the symbol of Minerva on the reverse. Lettsom wrote that "this medal, with the reverse, I design to ornament a new edition of my 'Observations on the Cow-Pock,' as exhibiting a patron of the great Jennerian discovery of Vaccination."[45] This relates directly to Lettsom's role as a champion of vaccination as a public health policy, and he first published his *Observations* in 1801 to promote Jenner's findings, just three years after Jenner's own publication announcing his discovery. Lettsom was also one of the members who established the Jennerian Society for the Extermination of the Small-pox.

The use of Minerva as presiding patron may also have been a reminder of the progress medical practice was making and Lettsom's own promotion of vaccination through both words and objects. The same circulation through networks as botanical works and specimens is reflected in the sending of what he called the first "Vaccine lymph" for smallpox to the United States via Dr. Waterhouse, professor of the theory and practice of medicine at Harvard Medical School, Massachusetts.[46] In 1805, writing to Plumtre, Lettsom discussed his hopes for the new vaccine, stating that "about three years ago, vaccination produced a sensible effect upon the annual deaths in London, which were 1,200, the year before last 1,100, and the last year 600 so that instead of 6,000 deaths, we experienced only 2,900. In Germany, vaccination has nearly extinguished the small-pox."[47] It would of course take until the mid-twentieth century for the global eradication of smallpox by the World Health Organization to be a reality, but Lettsom was clearly progressive in his approach to medical practice.

His scientific interests also extended to the recording of weather conditions using a thermometer and barometer. A microfiche of a diary he kept in the last years of his life, 1812–1813, is retained in the Wellcome Collection and is full of climatic observations as well as peppered with notes on the lectures he gave, people he dined with, and, on one occasion, a journey taken by boat to Margate to check on the sea-bathing institution he had founded in 1791.[48] His climatic recordings are a reminder of the constant measurements of the natural world and the garden as a place of scientific observation. On June 1, 1812, he recorded: "The weather close and sultry, the Therm. 70 and the Barom 29, wind south east," and for November 4 and 5 he stated that they "were partly foggy but fine midday, south a bright moon. Ice was about the thickness of a half crown." Here he was using a thermometer and a barometer as a scientific method of recording the conditions, but this is still supplemented with a sensory observation of the weather as "sultry," and visual observations of the brightness of the moon and the thickness of the ice. In between the scientific descriptions we are sometimes treated to more lyrical accounts of his atmospheric environment. In March 1812, for example, he recorded that at the start of the month there were "clear nights bright moon, & starry firmament," which by the conclusion had turned into weather that "was warm and pleasant, with some little rain; vegetation is forward and leaves are everywhere bursting from the buds. Throughout the month the wind has been generally moderate, and as grateful and warm as are usual in May." There is little to link this meteorological fascination to his interest in medical practice, although he does mention at the end of 1812 that "the year has been in great measure amensurable to health; no epidemic having been prevalent not even the smallpox." The relationship of climate to disease has a long history in Western medical practice that can be traced back to classic texts, such as the Hippocratic treatise of *Airs, Waters, Places,* and the measurement of weather was a popular pursuit of the gentry from the seventeenth century onward.[49] Domestic spaces and gardens could then form the backdrop to a variety of scientific practices.

Other uses of scientific instruments are outlined in John Hunter's letters to Jenner, by which we discover that he sends at least two thermometers (Jenner breaks the first), and requests Jenner measure the internal temperature of hedgehogs during hibernation.[50] Both Hunter and Jenner were interested in how animals could survive long periods of time, seemingly asleep, and used invasive techniques to try to understand this process. This fascination with the activities of hedgehogs led Hunter to console Jenner, when he hears he has been unlucky in love, with the immortal phrase: "But 'let her go, never mind her.' I shall employ you with hedge hogs, for I do not know how far I may trust mine."[51] As well as raising the question of how hedgehogs in the vicinity of Earl's Court might be considered unreliable, presumably as scientific objects, this constant measurement of the world around them by Lettsom, Jenner, and Hunter fulfills Lorraine Daston's description of the development of "observation as a way of life" in this period.[52] Here the measurement of the air and the observation of the stars as well as the temperature of a hedgehog were all domestic activities that were practiced within gardens and recorded in diaries and letters.

There are other shared connections between Lettsom and Jenner. The use of the garden as both a rural idyll and a site of modern scientific and medical practice is perfectly illustrated by the use of The Chantry's gardens in Berkeley by Jenner. In 1933 the Wellcome Research Institution exhibited a series of dioramas as part of their showcase at the Chicago World Fair, which were intended to illustrate the history of medicine in line with the exposition's theme of "A Century of Progress." Among these was a simple rustic garden building used to depict one of the arguably greatest medical accomplishments—the introduction of vaccination as a public health method (fig. 6.5).

"I have given my little cottage the name of the Temple of Vaccinia": Medical Science and the Garden

Jenner's estate in Berkeley (fig. 6.6), on the edge of a small town in deepest Gloucestershire, was far more rural than Lettsom's semimetropolitan retreat, but its picturesque nature and setting for scientific and medical

Fig. 6.5. Diorama of Jenner's Temple of Vaccinia as displayed at the Chicago World Fair, "A Century of Progress," 1933. From the Wellcome Research Institution. Wellcome Collection, CC BY 4.0.

pursuits had much in common with the domestic research center at Grove Hill. Like the plant fertilizer experiments described in chapter 5, Jenner pursued a variety of interests relating to the natural world. In 1798 he was elected to become a fellow of the Linnean Society; he was a member of the Royal Geological Society and was conferred membership of the Royal Society for his observational work on cuckoos. Like many of the physicians we have considered in this book, he was described by his biographer, John Baron, in 1838 as having "knowledge of the economy of plants and animals" and paying "vigilant attention to all the varied forms and properties of surrounding objects."[53]

Undoubtedly the most interesting physical structure in Jenner's garden is the summerhouse known as the Temple of Vaccinia, in which Jenner conducted free vaccinations for the poor. On May 19, 1804, a Mr. Joyce

Fig. 6.6. Jenner's home and garden, at The Chantry, Berkeley, in Gloucestershire, with a greenhouse slightly hidden by shrubbery at the front right, as drawn by Stephen Jenner (probably Edward Jenner's nephew). Engraving published 1826. Print in author's own collection.

wrote to Lettsom and described his visit to Jenner. He arrived as the physician was sitting down to breakfast and recalled:

> This parlour in which we were sitting look'd into an agreeable lawn, on one side of which ran a walk. . . . I had observed during our conversation a great number of females with children in their arms or by their side, passing down this walk . . . ; and I could not forebear interrupting the conversation to enquire of my friend, what it meant. It has been the custom for some time, said he, to set apart one morning in the week for inoculating the poor.[54]

As well as describing the process by which people would arrive and queue to be vaccinated, Joyce continued to report the building as described to him by Jenner:

Fig. 6.7. Painting of Jenner sitting outside his Temple of Vaccinia. Date and artist unknown. Wellcome Collection, CC BY 4.0.

> In the midst of those trees is a small mansion built in the cottage stile. It consists of one room only and was erected for the purpose of giving a rural appearance to that part of my garden. I have lately converted it into a place of utility—and the people who come to be inoculated assemble there and wait until I come among them. It is for this reason, I have given my little cottage the name of the Temple of Vaccinia.[55]

This makes it clear that this building was originally ornamental and later adapted in its use, and that the name Temple of Vaccinia was conferred upon it by Jenner before 1804.[56]

The original temple, then, was simply a summerhouse in the picturesque style—one that appears to fit the eighteenth-century aesthetic of a rustic retreat that could be used for contemplation (fig. 6.7). This type of construction was illustrated in Thomas Wright's *Arbours* and *Grottos,* published in the 1750s. These two volumes contained Wright's designs for garden retreats. The rustic style is most evident in the *Arbours* volume, with designs "for a Hut or Hovel-kind, chiefly designed for a shelter'd solitude" and "a Druid's Cell, or Arbour of the Hermitage Kind, purposely designed for a Study or Philosophical retirement."[57] This is not to suggest that Wright had anything to do with Jenner's building, but the rusticated style would be associated by contemporaries with activities such as contemplation. As Gervase Jackson-Stops established, hermitages were "always primitive and rustic, made of boulders, roots or bark, and with thatched or turfed roofs, [and] just as popular with gentry of more modest means."[58]

A similar hut designated as a "hermitage" was built by the natural historian and writer Gilbert White of Selborne in the 1750s, which provides further evidence for the relationship between the rustic style and its use as a retreat. As this was depicted in paintings—including a 1777 watercolor by Samuel Hieronymus Grimm with Henry White, Gilbert's brother, dressed as the hermit—this rustic building might have provided some inspiration, particularly given the mutual interest of Jenner and White in natural history (fig. 6.8).[59] Jenner's temple also bears a striking resemblance to a hermitage constructed by Matthew Boulton, engineer and manufacturer, on his Soho estate in Birmingham (fig. 6.9). Both White and Boulton had scientific and natural history interests, and this suggests that Jenner can be viewed on one level as a romantic gentleman who used his garden foremost to display his aesthetic taste, in a similar manner to Lettsom. This space was then adapted and clinically appropriated for the performance of vaccinations.

The shift from its use as a rural retreat to a key site of public health care also denotes a broader move for medical practitioners away from pastoral pastimes to a professional status that revolved around modern medical practices. The secondary use of the temple as the space in which Jenner vaccinated the poor against smallpox has led to the building's greater historical significance. The reception of the garden building as a place of medical importance has developed out of Jenner's seminal work on smallpox and his dissemination of the vaccination methodology as a successful form of preventive medicine (fig. 6.10).

Moving into the Victorian period, the lives of medical practitioners were transformed. Jenner's Temple of Vaccinia signals this gradual shift

Fig. 6.8. Detail from title page of Gilbert White's *Natural History of Selborne.* White's brother is dressed as a hermit standing outside a small straw-covered, conical hermitage. After Grimm (London: Printed for White, Cochrane, 1813). Wellcome Collection, CC BY 4.0.

toward a medical profession that by the mid-nineteenth century was built on clinical experience, medical knowledge, and competence rather than other forms of polite knowledge.[60] As Brown notes, these later men "owed their reputations to their abilities as clinicians. They felt pulses and listened to heartbeats. They did not generally write books about trees."[61] This is not to say that medical practitioners no longer had an interest in botany, agriculture, or horticulture, but that their estates no longer held the same significance as an entrance ticket to the status of medico-gentility.

This does not imply that subjects like botany immediately stopped being relevant or considered an important element of medical education. In 1829, Joseph Houlton, general practitioner and fellow of the Royal Medico-Botanical Society, wrote to *The Lancet* requesting subscribers for a new botanic garden in northwest London for the use of medical practitioners.[62]

Fig. 6.9. Matthew Boulton's Hermitage at Soho House, Birmingham, which is similar in style to Jenner's Temple of Vaccinia at Berkeley. Photo by author, 2019.

He argued that this was an essential undertaking, as "medical botany is now made a branch of medical education, and that there is no public collection of medicinal plants near town."[63] However, it is clear from the 1830s onward that the role of botany as a key part of medical education was under threat. Writing in 1837, William Howison, lecturer in botany at Edinburgh, bemoaned the replacement of botany from the curriculum of the Royal College of Surgeons in Edinburgh with mathematics and mechanical philosophy. This represented a move away from the more traditional curriculum, still including botany, that Howison was involved with teaching at the University of Edinburgh.[64] Harkening back to earlier times, he also praised the

Fig. 6.10. Restored Temple of Vaccinia at Berkeley. Photo by author, 2013.

use of herborizing as the only real method of botanical teaching and was critical of the use of botanic gardens with their cultivated species.[65] He also lamented the use of botanical models for teaching, complaining that on a trip to an institution in Europe they had shown him "a plant made of *sheet iron,* with its flowers, leaves &c., painted to resemble nature, and this was the principal means of teaching his students! Shade of Linnaeus! couldst thou have entered this man's lecture room, what wouldst thou have said at witnessing thy favourite study so prostituted and abused?"[66] This underlines the still-growing importance of the use of objects in medical teaching, in itself related to the long history of the use of medical museums within medical education.[67] The garden itself, however, gradually seems to lose its importance as a scientific tool, as the hospital and eventually the laboratory became more central.

Frustrations with the multiplicity of subjects medical students were expected to undertake as part of their studies had been voiced by as early as

1845 by G. B. Knowle, professor of botany and materia medica at Queen's College, Birmingham. He still argued for a central role of botany as an important subject, as it taught students close observation of natural structures, but it is clear that by the mid-nineteenth century, the extensive medical curriculum, with its focus on hospital-based training, gradually eased out botany as a key subject.[68] By the 1870s, F. William Headland, physician to Charing Cross Hospital, was arguing that elements such as botany should be taught before students arrived at medical schools. This was, he argued, so that students would have more time to devote themselves to "those practical studies in the hospital which are by far the most important part of his medical education."[69] Again this emphasizes the shift to clinical medicine as a keystone of the curriculum.

Other associated cultural and social changes also meant that there was a move to new urban sensibilities that guided the accessibility and placement of many collections. At the start of the nineteenth century the movement to more publicly accessible spaces for museums, as well as libraries and gardens, reflected new notions of what was public or private in relation to the home. This also encouraged a growth in the collective ownership of objects and spaces. The core of many more public collections was formed from personal acquisitions established in the intervening centuries. This transformation, traced by Sam Alberti in relation to museums, was also true for some botanic collections. The creation of the Glasgow Botanic Garden in the 1810s was driven by Glasgow citizens led by the wealthy Thomas Hopkirk, with joint funding from subscribers, through the Royal Botanic Institution of Glasgow and the University of Glasgow.[70] The 1818 guidebook remarked that as well as donations of plants from other botanic gardens in Dublin, Edinburgh, and Liverpool, private donors were also key. In particular they mention Hopkirk by name, whose "whole private collection consisting chiefly of hardy plants, was transferred bodily from Dalbeth."[71] In this way private botanic collections were also moved and reused in different, more public spaces along with plants from other institutional collections.

The Afterlives of Gardens and Collections

In 1805, one year after the lavish party described at the start of this chapter, "Lettsom was complaining that he was 'right worn down with fatigue' and that Grove Hill was so inundated with visitors that it resembled a hotel."[72] By 1811 Nichols notes that Lettsom had "been compelled, by a train of adverse circumstances, at an advanced period of life, to dispose of the greatest part of so valuable a collection, and even of the Villa itself."[73] As he states, "One part of the Library was sold, March 26, 1811, and six following days, by Messrs. Leigh and Sotheby; by whom the remaining part was also sold, April 3–5; and the entire Museum, including Coins and Medals, May 2–4, 1816."[74] Having sold his villa, Lettsom spent the final years of his life at his central London home at Sambrook Court, dying there in 1815. Like Fothergill, there was sadly to be no rural retirement to be enjoyed at the end of his working life.

There is very little to mark Lettsom's estate in Camberwell now, as the house was demolished in the 1890s, but a small patch of the garden ground survives and is currently managed by the charitable Lettsom Gardens Association; it is still known as Lettsom's Gardens (fig. 6.11). Similar fates befell John Hunter's estate at Earl's Court and Pitcairn's Islington garden. Hunter's house was demolished in the 1880s, and the landscape was built over with new suburban villas by Robert and John Gunter. Pitcairn's botanic garden is now subsumed by the development of Almeida Street, Upper Street, and Bathshill Street.[75] Similarly, having had several locations in South London, the London Botanic Garden's final site in Brompton was turned into a nursery in the 1820s and subsequently lost, and the only survivor of the Leith Walk garden, the botanic cottage, has recently been removed brick by brick from its original site and rebuilt in the current Royal Botanic Garden Edinburgh (plate 31).

Other sites owned by our medical practitioners have been somewhat more fortunate. Fothergill's garden at Upton is now West Ham Park; some vestiges of the landscape remain for those who look, and there are interpretation boards that at least offer the casual visitor some sense of the landscape's eighteenth-century history. The house itself, renamed as Ham

Fig. 6.11. What remains of Lettsom's Grove Hill now forms the basis of a community garden in Camberwell named Lettsom's Gardens. Photo by author, 2018.

House, was, however, demolished in 1872. The only remaining house and garden that has featured prominently in this work is Edward Jenner's garden at The Chantry in Berkeley, which, being a rural estate without the pressures of suburban metropolitan building and with a particularly prestigious former owner, has survived intact. Today the museum and garden are managed by the charitable Jenner Trust and, with its recently restored Temple of Vaccinia, can be visited by the public for a small entrance fee.

As with the houses and the gardens, the associated collections have also often been dispersed, sold, and are almost impossible to trace, the main exception to this being John Hunter's comparative-anatomy collection, which now forms the basis of the museum of the Royal College of Surgeons. Occasionally, individual specimens and grouped artefacts can be found subsumed within other collections. For example, Fothergill's shells and corals, which were obtained by William Hunter, still form part of the

Hunterian collection in Glasgow.[76] However, the role of collecting, which seems to have formed an element of everyday life for Georgian medical practitioners, is hard to reconstruct without the physical objects or buildings and landscapes in which they were located.

We are left, then, with little tangible evidence that we can hold or landscapes that we can visit which would directly connect us physically to these past landscapes and collections. However, as this research has revealed, they once formed part of personal, local, national, and global networks and were visited and studied by numerous people. They were the material expressions of a time when the gentleman medical practitioner was also an experimental, botanic, scientific, and picturesque landscaper. By analyzing these multilayered landscapes belonging to an emerging professional class, we can also offer a window into the past for the modern-day garden visitor, whose own garden spaces are often experimental and combine the utilitarian with the beautiful.

EPILOGUE

The Stories We Tell: Bridging the Gap between Research and Practice

GIVEN THAT FEW OF THE original landscapes discussed in this book remain in anything like their original state, it is evident that in practical terms alone most of these landscapes are not accessible to today's garden-visiting public (Jenner's garden in Berkeley and West Ham Park on the site of Fothergill's landscape being notable exceptions). However, as interest grows in how British landscapes can be made more inclusive for visitors and how gardens can be interpreted in more creative ways, a complementary research approach that expands beyond the visual design to consider the multisensorial experience as well as the historic use of such spaces is crucial.

In this epilogue, I argue that the heritage sector, which is developing exciting and more inclusive approaches to interpretation, can only be enhanced by historical work, which unpacks our cultural assumptions and moves beyond traditional garden history approaches. Although this book by its nature represents a very one-sided academic approach to this problem, I argue that ideally this process should be a two-way movement in which research questions are opened and shaped by practical interpretation and engagement goals, rather than being seen as two separate, albeit interlinked, enterprises.

This is, of course, not a new concern. Back in 2007, Williamson and others were already arguing that garden history as a discipline was still predominately tied to art and literary history methodologies that prioritized certain ways of looking and thinking about landscapes.[1] In particular, Williamson noted that researchers tended to overlook areas such as "the complex ways in which gardens were used by contemporaries, as arenas of social display and recreation, and as statements of social identity—and how these things helped to shape their form."[2] This was picked up again in 2013 in a report conducted by Gregory, Spooner, and Williamson for the national heritage organization Historic England, wherein they noted that even with well-known gardens created by Lancelot "Capability" Brown, more research was essential. In particular, they state that researchers rarely consider how such parks were "used and experienced" and how elements such as gender affected how such landscapes were perceived and understood.[3] Such calls to highlight the use and experience of gardens have clearly been made, however garden history has generally remained more concerned with the material, and particularly the visual, nature of the landscape, than how it was experienced by people in the past.

Engaging Diverse Audiences

As John Wylie has written, research on landscapes to date has generally focused on a "particular *visual* mode of observing and knowing," which is captured by the concept of "reading the garden," whether by viewing it as a document to be interpreted or by walking through it as a methodology.[4] This visual approach, which is embedded in much earlier debates regarding concepts such as the picturesque, sublime, and even beauty in relation to landscape, can unintentionally exclude those with different cultural perspectives. Similarly, this approach can close off avenues of wider historical exploration when applied to interpretation schemes that unintentionally alienate those who experience landscape primarily through sound, touch, scent, and movement.[5] Shifts in this area are occurring, though. Spearheading this approach is a new publication from Dumbarton Oaks, *Sound and Scent in the Garden,* edited by D. Fairchild Ruggles, which has opened the door to a broader academic conversation about the sensory history of a range of landscapes from around the globe.[6] Botanic gardens have already been leading the way, with places such as Oxford Botanic Garden encouraging visitors to engage sensorially with select plants in their collection (plate 32) and a similar concept promoted by the Royal Botanical Gardens,

Fig. 7.1. The successful theme of the Royal Botanical Gardens in Canada in summer 2019 was "Come to Your Senses," and visitors were encouraged to engage on a multi-sensory level with a range of plants. Photo by author, 2019.

Canada with their 2019 theme of "Come to Your Senses" (fig. 7.1). I would argue that these are exciting approaches, but that fruitful new avenues of research might be found by considering historical approaches, such as the history of science, sensory and emotional history, *in conjunction* with new forms of heritage interpretation.

Changing areas of focus within the historical academic community also offer expanding approaches to the subject. For instance, the developing interest of environmental historians in gardens as places of study has already been touched on in my introduction and will hopefully become a growing area for future researchers.[7] Leaping the fence between disciplines is just one approach, which can only enhance the ways in which we research and interpret gardens. The call to decolonize the history curriculum in a recent report by the Royal Historical Society, for example, offers a valuable opportunity to engage students with histories of plants, gardens, and gardening as seen through a colonial lens.[8] There is also an opportunity here to link up with other heritage and cultural organizations that are grappling with similar problems. As the eighteenth-century garden was interlinked with museums, libraries, and other collections, by looking to the past we can also see new directions for considering the threads that link contemporary collections together.

These approaches should also resonate with current concerns of garden visitors and may open up new spaces for discussion. To some extent the Eden Project in Cornwall (opened in 2001) offers a model for engaging people with plants in relation to wider issues beyond the immediate garden, and one that applies diverse approaches that move beyond the purely aesthetic and encourage alternative forms of engagement. Although the assertion of Smit and Kendle in 2011 that "for the wider public, botanic gardens had become irrelevant to their lives" feels unjustified, given the visitor numbers still attracted to places such as Kew and Edinburgh, the Eden Project has tackled important current challenges, such as climate change, food security, and the hidden agricultural labor underpinning our supermarket produce.[9] In many ways Eden, with its mission of "exploring how we can work towards a better future," is a contemporary version of Lettsom's Camberwell garden, which aimed to improve agricultural and botanical knowledge necessary for the challenges of his age. The development of elements such as social-prescribing projects at Eden also denotes a new and important direction toward more inclusive approaches to gardens and gardening.[10] This is further reflected in the work of the Sensory Trust and Historic England, which highlights alternative ways of thinking about how diverse groups of people interact with and respond to landscape.[11]

It is clear that the role of science in the garden also has the potential to be used to engage audiences with current and future concerns, particularly in relation to food and agricultural experimentation. By recognizing the garden as a laboratory space in the past, with its cultivation of new exotic and utilitarian species, the historic landscape offers a place for scientific debate and exploration of modern issues. Similarly, the collection of other forms of data related to weather and climate in historic gardens, as described by Alexandra Harris in *Weatherland* and Mark Laird in *A Natural History of English Gardening,* can also lead to the garden being utilized as a space in which to discuss our changing climate.[12] Gardeners are often already aware of the role of climate in their own backyards, and the measurement of weather conditions recorded by earlier men like Lettsom could provide a vital link to past concerns, such as the knowledge of when the last frost is likely to occur, and the approaches of acclimatization and experimentation taken by gardeners in the past. It can also encourage people to consider the global scale of challenges faced by those working with plants and the diverse challenges faced by those in other more precarious geographical locations than the United Kingdom. Again, the Historic England report on Brown noted the importance of considering climate change in relation to management and sustainability in the future, but such issues could also lead to more conversations between garden visitors and those working on future-proofing landscapes now.[13]

In particular, by underscoring the networks between and beyond gardens, we can start to include those voices silenced by empire and consider the ways in which slavery and the eclipsing of indigenous knowledge have been erased from our understanding of histories of garden creation, collecting, and display. This is already being explored by the outstanding Colonial Countryside project led by Corinne Fowler at the University of Leicester in conjunction with the National Trust. This is a truly exciting example of how we might bring these narratives into conversation with the traditional histories of country-house estates and make them visible to visitors.[14] Building on this, and responding to the powerful and influential global Black Lives Matter protests in the summer of 2020, the National Trust has also published its pioneering report *Connections between Colonialism and Properties Now in the Care of the National Trust,* which describes and highlights the interconnections between slavery, empire, landowners, and heritage.[15] Of course, once one has read Tobin's groundbreaking work, which reframes the idea of the "exotic" plants in the British garden in relation to the colonial conquest of the tropics and the associated elision of both indigenous labor and knowledge, it is difficult to see the "English" landscape without also considering it as an artifact of power and exploitation.[16] As such attention has finally forced us to really see this relationship, it has led to organizations such as the Royal Botanic Gardens, Kew, seeking to decolonize their collections.[17] The political events of 2020 finally appear to be shifting this research landscape, and I can only hope that by the time this book is published, numerous projects and doctoral scholarships will have emerged that eventually redress this lacuna in our understanding and encourage a broader appreciation of the location of gardens within the web of empire and colonialism.

Storytelling

Moving toward these more inclusive approaches to studying gardens should in turn extend to the development of new stories about our historic landscapes. In *The Wilderness* podcast, Jon Favreau interviewed Barack Obama about the importance of storytelling in speechwriting. Obama summed up his approach by stating that "in simple terms, people learn from stories."[18] This is a theme that is currently of interest to historians, with projects such as the *Storying the Past* reading group and the associated Creative Histories events, organized by Will Pooley, University of Bristol.[19]

As landscape historians, our research can change the stories we tell, which will not only help people learn about the broader histories of the landscapes themselves, but also have the potential to be more inclusive in the narratives chosen. As Steve Poole notes in relation to his groundbreaking Ghosts in the Garden project, their intention was to

> suggest to visitors that a place has many histories and that our understanding of it is influenced by a process of narrative selection. The essential proposition was that quotidian stories and characters from the historical record can be as engaging to audiences as stories about celebrities and social elites because they reflect more closely the life experiences of modern garden visitors.[20]

The multinarratives in this book—drawing on the involvement of garden visitors, including women, and gardeners as expert technicians, as well as connections to wider histories—allow for richer interpretive experiences for today's garden visitors. Women in particular are being brought into focus in relation to gardens more strongly, with Catherine Horwood's *Gardening Women;* Madeleine Pelling's work on Margaret Cavendish Bentinck, the Duchess of Portland, at Bulstrode; and a 2019 special issue of the Women's History Network journal dedicated to gardening, offering just a few recent and welcome examples.[21]

Although the issue remains that with less elite women and the laboring classes, the limited nature of primary source material makes reconstructing their impact a challenge. As Carole O'Reilly notes in her recent work on parks, "Any discussion of the staff of the public park is hampered by . . . a corresponding lack of material emanating from the gardener and the labourer. The accounts of park-keepers are similarly rare. Thus, any attempt to tell the story of the parks' employee is limited."[22] However, even limited snapshots that acknowledge the often-hidden labor in the making, remaking, and maintenance of landscapes would help raise the significance of such roles within landscape history.

A concern with the close relationship between research and interpretation was key to Poole's choose-your-own-adventure project developed in conjunction with an experience design company, which used the Sydney Gardens in Bath as its setting. As Poole argues, "The question we should in any case be asking perhaps, is not 'how can we use mobile and digital technologies to get larger and more diverse audiences through the door?,' but 'how can we use mobile and digitally enhanced forms of interpretation to change the questions we ask and the ways in which we engage with historic sites?'"[23] It is this interrelationship between interpretation and research that can bridge research and practice and also offer novel and timely questions and approaches for researchers to investigate. There is a hint here of the range of new research possibilities offered by digital technologies, with the caveat that this should enhance the findings of the academic research rather than be seen as the solution to inclusivity in itself.

In this way we can use the garden to talk about the more difficult histories of the past. As stories that are more commonly "erased," these histories range from the economic funding of gardens via the slave trade (as highlighted by University College London and the University of the West of England with Historic England, on their different projects linking slave ownership to broader data, including the financing of country houses and their estates) to the sweeping away of villages for aesthetic reasons in the eighteenth century.[24] This dark heritage can surely be explored more widely through historical interpretation that extends beyond the visual reading of the landscape, suggesting a new range of lenses by which the past can be explored.

In the United Kingdom the recent Capability Brown and Humphry Repton celebratory festivals, although seemingly traditional in their focus on great male landscape architects and their elite garden designs, have encouraged a range of novel activities aimed at developing new audiences. Examples include the Potter and Ponder collaboration at Croome, which developed new sensory interpretation techniques, and the Garden Trust's Sharing Repton project, which piloted activities designed to help volunteers welcome wider local communities to landscapes designed by Repton and then share their learning experiences with others.[25]

All such work points toward a trend of new and exciting approaches to landscape engagement, storytelling, and interpretation, which could be further enhanced if underpinned and informed by historically sensitive academic research. Because garden history as a defined subject has declined as a dedicated academic subject area in the United Kingdom, there has naturally been concern raised in relation to expertise within the academy. However, the growing interest in environmental and sensory history, experience, and use as categories of research, and the associated spatial and material turns more broadly, should hopefully open up new and exciting avenues of inquiry and set landscape studies far beyond the well-trodden garden path, to a bold new future.[26]

Notes

Introduction

1. Lettsom, *Grove-Hill: An Horticultural Sketch.*
2. Pettigrew, *Memoirs of the Life and Writings of John Coakley Lettsom,* 1:32.
3. The concept of centers of calculation was first posited by Bruno Latour in *Science in Action* (Cambridge, MA: Harvard University Press, 1987), and the importance of paying attention to other places of scientific calculation has been demonstrated by Stewart in "Other Centres of Calculation."
4. For more on Joseph Banks and Kew, see Gascoigne, *Joseph Banks and the English Enlightenment.* The use of other networks and garden spaces as experimental places and their roles in the shaping of gentlemanly status in the early eighteenth century has been explored by Coulton in "Curiosity, Commerce and Conversation in the Writing of London Horticulturalists."
5. Joseph Levine, "John Evelyn: Between the Ancients and the Moderns," in O'Malley and Wolschke-Bulmahn, *John Evelyn's "Elysium Britannicum" and European Gardening,* 69.
6. Classen, "Museum Manners."
7. Hunting, "Dr John Coakley Lettsom, Plant-Collector of Camberwell."
8. Symes, *The Picturesque and the Later Georgian Garden,* 9–10.
9. Ibid.
10. Hunting, "Dr John Coakley Lettsom, Plant-Collector of Camberwell"; and Clark, "William Curtis's London Botanic Gardens."
11. Anonymous, "John Coakley Lettsom MD."
12. Ibid., 470.
13. By 1800, his practice reaped an incredible annual income of around £12,000. J. F. Payne and Roy Porter, "Lettsom, John Coakley (1744–1815), Physician and Philanthropist," *Oxford Dictionary of National Biography* (Oxford: Oxford University Press, September 23, 2004).
14. Spooner, *Regions and Designed Landscapes.* See also Rebecca Preston, "Home Landscapes: Amateur Gardening & Popular Horticulture in the Making of Personal, National and Imperial Identities, 1815–1914" (unpublished PhD thesis, Royal Holloway, University of London, 1999).
15. Laird, *Flowering of the English Landscape Garden,* 158.

16. Stobart, "'So Agreeable and Suitable a Place,'" 89.
17. Ibid.
18. Brockway, "Science and Colonial Expansion."
19. Drayton, *Nature's Government,* 89.
20. Brown, *Performing Medicine.*
21. The recent tercentenary celebrations for Capability Brown have resulted in a flurry of books dedicated to Britain's most well-known garden designer, including but not limited to Sarah Rutherford, *Capability Brown and His Landscape Gardens* (London: National Trust, 2016); David Brown and Tom Williamson, *Lancelot Brown and the Capability Men: Landscape Revolution in Eighteenth-Century England* (London: Reaktion Books, 2016); Steffie Shields, *Moving Heaven and Earth: Capability Brown's Gift of Landscape* (London: Unicorn, 2016); Phibbs, *Place-Making.* Other works on key designers include Stephen Daniels, *Humphry Repton: Landscape Gardening and the Geography of Georgian England* (New Haven, CT: Yale University Press, 1999); Fiona Cowell, *Richard Woods (1715–1793): Master of the Pleasure Garden* (Woodbridge, UK; Rochester, NY: Boydell Press, 2009). Recent books by time period include the excellent volume by Patricia Skinner and Theresa Tyers, eds., *The Medieval and Early Modern Garden in Britain: Enclosure and Transformation, c. 1200–1750* (London: Routledge, 2018); and key examples of recent work organized by region include the Historic Landscapes of England series of county guides by Timothy Mowl and his team of researchers, and Tom Williamson's work with others, including the recent volumes describing the work of Humphry Repton in Hertfordshire (Hatfield, UK: Hertfordshire University Press, 2018) and in Norfolk (Norwich: Norfolk Gardens Trust, 2018).
22. Spooner, *Regions and Designed Landscapes,* 17.
23. Notable examples of such work include Easterby-Smith, *Cultivating Commerce;* Margaret Willes, *The Gardens of the British Working Class* (New Haven, CT: Yale University Press, 2014); Helena Chance, *"The Factory in a Garden": A History of Corporate Landscapes from the Industrial to the Digital Age* (Manchester, UK: Manchester University Press, 2017); Hazel Conway, *People's Parks: The Design and Development of Victorian Parks in Britain* (Cambridge: Cambridge University Press, 1991); Fiona Fisher and Rebecca Preston, "Light, Airy and Open: The Design and Use of the Suburban Public-House Garden in England between the Wars," *Studies in the History of Gardens and Designed Landscapes* 39, no. 1 (2019); Jane Hamlett, Lesley Hoskins, and Rebecca Preston, eds., *Residential Institutions in Britain, 1725–1970: Inmates and Environments* (London: Pickering & Chatto, 2013); Hickman, *Therapeutic Landscapes;* Katy Layton-Jones, *Places of Health and Amuse-*

ment: Liverpool's Historic Parks and Gardens (Swindon, UK: English Heritage, 2008); and O'Reilly, *The Greening of the City.*

24. Malcolm Dick and Elaine Mitchell, "Introduction: Gardens and Green Spaces in the West Midlands since 1700," in Dick and Mitchell, *Gardens and Green Spaces,* 1.
25. Ibid.
26. See note 23 above for key examples.
27. Williamson, *Polite Landscapes,* 4–5.
28. Easterby-Smith, *Cultivating Commerce.*
29. Felus, *The Secret Life of the Georgian Garden,* 5.
30. A forthcoming special issue of the *Journal of the History of Collections* will contain contributions from a sister American conference at The Huntington Library, Art Museum, and Botanical Gardens held in September 2017, on the same theme, articles edited by Anne Goldgar and Miles Ogborn.
31. Findlen, *Possessing Nature;* MacGregor, *Curiosity and Enlightenment.*
32. Beattie and Jones, "Editorial."
33. Laird, *A Natural History of English Gardening,* 5.
34. Fischer, Remmert, and Wolschke-Bulmahn, *Gardens, Knowledge and the Sciences.*
35. Findlen, *Possessing Nature,* 6.
36. Easterby-Smith, *Cultivating Commerce.*
37. Livingstone, *Putting Science in Its Place.* See also Livingstone, "Keeping Knowledge in Site," 782; Livingstone and Withers, *Geographies of Nineteenth-Century Science;* Elliott, *Enlightenment, Modernity and Science,* 12; and Elliott, Watkins, and Daniels, " 'Combining Science with Recreation and Pleasure.' "
38. Spary, *Utopia's Garden.*
39. Therese O'Malley, "Art and Science in the Design of Botanic Gardens, 1730–1830," in Hunt, *Garden History;* Johnson, *Nature Displaced, Nature Displayed.*
40. Exemplary examples of this are Kohler, *Landscapes and Labscapes;* and Naylor, *Regionalizing Science.*
41. Opitz, Bergwik, and Van Tiggelen, *Domesticity in the Making of Modern Science.*
42. Schiebinger, *Plants and Empire,* 10.
43. Tobin, *Colonizing Nature,* 21.
44. Ibid.; Drayton, *Nature's Government;* Miles Ogborn, "Vegetable Empires," in Curry et al., *Worlds of Natural History.*
45. For more on this, see Roy Porter, *Health for Sale: Quackery in England, 1660–1850* (Manchester, UK: Manchester University Press, 1989).

Chapter 1. Educating the Senses

1. Pettigrew, *Memoirs of the Life.*
2. Ibid., 17.
3. Ibid., 17–18.
4. Minter, *The Apothecaries' Garden.*
5. Ibid.
6. Semple, *Memoirs of the Botanic Garden at Chelsea,* 137–38.
7. Ibid., 140.
8. Noltie, *John Hope,* 19.
9. Rosner, *Medical Education in the Age of Improvement,* 63.
10. Ibid., 56.
11. As quoted and discussed by Rosner in ibid., 57.
12. Anonymous student, *Lectures on Botany by John Hope, 1777–8* (Edinburgh: Royal Botanic Garden Edinburgh [hereafter RBGE] Archives), 3.
13. Dugan, *The Ephemeral History of Perfume,* 156. See also William Tullett on the sensory pleasure garden in "The Macaroni's 'Ambrosial Essences': Perfume, Identity and Public Space in Eighteenth-Century England," *Journal for Eighteenth-Century Studies* 38, no. 2 (2015).
14. Anne Vila, "Introduction: Powers, Pleasures and Perils," in Vila, *A Cultural History of the Senses,* 4.
15. Michel Baridon, *A History of the Gardens of Versailles* (Philadelphia: University of Pennsylvania Press, 2008).
16. Vila, "Introduction," 7.
17. Mark Jenner, "Tasting Lichfield, Touching China: Sir John Floyer's Senses," *Historical Journal* 53 (2010): 653.
18. Ibid., 648.
19. Ibid., 654.
20. Ibid., 647–48.
21. Ibid.
22. Anonymous student, *Lectures on Botany by John Hope,* 13–14.
23. Dugan, *The Ephemeral History of Perfume,* 185.
24. Ibid., 182.
25. Cullen, *Lectures on the Materia Medica,* 1–2.
26. Ibid., 136.
27. Ibid., 137.
28. Ibid.
29. Quoted in Michael Bull, Les Back, and David Howes, *The Auditory Culture Reader* (London: Bloomsbury, 2015), 31.

30. Quoted in Vila, *Cultural History of the Senses,* 11.
31. Cullen, *Lectures on the Materia Medica,* 1.
32. Noltie, *John Hope,* 58.
33. For a detailed description of Hope's lectures, see ibid., 58–83.
34. Findlen, *Possessing Nature,* 257.
35. John Dixon Hunt, "The Botanical Garden, the Arboretum and the Cabinet of Curiosities," in Hunt, *A World of Gardens.* As John Dixon Hunt has identified, the gardens at Leiden and Uppsala were the work of distinguished botanists Clusius and Linnaeus, respectively.
36. Ibid., 131.
37. Physic gardens for training medical students and apothecaries became popular after the establishment of Padua in 1545, appearing in Rome in 1566, Zurich in 1561, Lyons in 1564, Bologna in 1567, Montpellier in 1598, Paris in 1640, the University of Oxford in 1621, and the Chelsea Physic Garden in 1673. See Johnson, *Nature Displaced, Nature Displayed,* 3.
38. Ibid.
39. Hunt, *A World of Gardens,* 133.
40. Findlen, *Possessing Nature,* 257.
41. Ibid.
42. Ibid.
43. Johnson, *Guide for Gentlemen Studying Medicine,* 14. For more detail on this guide and its authorship, see Rosner, *Medical Education in the Age of Improvement.*
44. Ibid.
45. Ibid.
46. Brown, *Performing Medicine.*
47. RBGE Archives, GD/253/144/8; Noltie, *John Hope,* 86.
48. Noltie, *John Hope,* 86.
49. Rosner, *Medical Education in the Age of Improvement,* 63.
50. October 20, 1807, Minutes of Meetings of the Faculty, 1806–1813, University of Glasgow Archives 26697, Clerk's Press 82.
51. Hill, *An Idea of a Botanical Garden,* 8.
52. J. R. Sealy, "Royal Botanic Gardens at Kew," *Science,* new series, 129, no. 3360 (1959): 1403.
53. More detail on John Hope and the Leith Walk garden can be found in Noltie, *John Hope,* and in particular 84.
54. Ibid., 18. Noltie also notes that when Hope traveled to Europe as a medical student in the 1740s, he chose Paris over Leiden because he wished to study under de Jussieu because of his interest in botany, ibid., 9. See Spary,

Utopia's Garden for a detailed and enlightening exploration of the various political, social, and scientific roles played by the Jardin du Roi, Paris.

55. Therese O'Malley, "Art and Science in the Design of Botanic Gardens, 1730–1830," in Hunt, *Garden History,* 286.
56. Lausen-Higgins, "Sylva Botanica."
57. O'Malley, "Art and Science," 284.
58. Ibid., 294.
59. See ibid.
60. Koerner, *Linnaeus,* 14.
61. Noltie, *John Hope,* 54–55.
62. Koerner goes so far as to argue that "Linnaeus' binomials resulted from his attempts to practice science as an auxiliary branch of economics, and from this efforts to create a simple language for it." *Linnaeus,* 43.
63. Arnot, *History of Edinburgh,* 402.
64. O'Malley, "Art and Science," 299.
65. Anonymous student, *Lectures on Botany by John Hope,* 3.
66. Gerrit Arie Lindeboom, *Boerhaave and Great Britain: Three Lectures on Boerhaave with Particular Reference to His Relations with Great Britain* (Leiden: Brill, 1974), 26.
67. Ibid., 19.
68. Derek Doyle, "Edinburgh Doctors and Their Physic Gardens," *Journal of Royal College of Physicians of Edinburgh* 38 (2008): 365.
69. Ibid.
70. Noltie, *John Hope,* 24–25.
71. The cause of Williamson's death was related to his second employment as a customs officer. It was while acting in this capacity that he was mortally wounded by a group of armed smugglers. Ibid., 36.
72. Stephen Harris, *Oxford Botanic Garden & Arboretum: A Brief History* (Oxford: Bodleian Library, 2017), ix.
73. Anonymous student, *Lectures on Botany by John Hope,* 7.
74. *A List of Specimens of Experiments Kept in the Gardeners House,* Botanical Papers of John Hope MD Professor of Botany and Materia Medica at Edinburgh, National Records of Scotland, copy read in the RBGE Archives, GD253/145/7/2.
75. Ibid.; Steven Shapin, "Invisible Technicians."
76. Noltie, *John Hope,* 36.
77. Jane Corrie, *Botanic Cottage Project Report. Stories from the Historical Archives: about Botanic Cottage, the Leith Walk Garden and John Hope's "Other" Life as a Physician,* May 2009 (RBGE/3B Cor).
78. Ibid.

79. John Hope, *Remarks on Lectures,* RBGE Archives, GD253/144/14/16.
80. Ibid.
81. Boney, *The Lost Gardens of Glasgow University.*
82. *Memorial by Robert Hamilton, Professor of Anatomy and Botany, and Dr William Cullen, Professor of Medicine, to the University concerning the Planting of More Trees and Shrubs in the College Garden in Place of the Decayed Fruit Trees,* University of Glasgow Archives: GUA 5412.
83. Ibid.
84. Letter from Dr. Brown to Dr. Jeffray, Thursday, June 12, 1806, University of Glasgow Archives: GUA 1961.
85. Ibid.
86. William Lang, "Representation by William Lang, College Gardener, to the Faculty Regarding Their Complaints about His Conduct and Neglect of College Garden," January 25, 1807, University of Glasgow Archives: GUA 1961.a.
87. For an expert reading of Rousseau and botany, see Cook, *Jean-Jacques Rousseau and Botany.*
88. Easterby-Smith, "Selling Beautiful Knowledge," 532.
99. Ibid.
90. From Berkowitz, *Charles Bell and the Anatomy of Reform,* 48.
91. The link between anatomical and botanical practices is generally overlooked when considering Georgian medicine, however the close relationship of the two subjects in the sixteenth century has been expertly described by Kusukawa in *Picturing the Book of Nature.*
92. Noltie, *John Hope,* 38.
93. *Caledonian Mercury,* March 4, 1776.
94. Ibid.
95. Noltie, *John Hope,* 38.
96. John Hope, "Remarks on Botanical Lectures," RBGE Archives, GD253/144/14/6.
97. Jo Currie, "Fyfe, Andrew (1752–1824), Anatomist," *Oxford Dictionary of National Biography* (Oxford: Oxford University Press, September 23, 2004).
98. Johnson, *Guide for Gentlemen,* 9–10.
99. Matthew Daniel Eddy, "Useful Pictures: Joseph Black and the Graphic Culture of Experimentation," in Robert G. W. Anderson, ed. *Cradle of Chemistry: The Early Years of Chemistry at the University of Edinburgh* (Edinburgh: John MacDonald, 2015).
100. Carin Berkowitz, "Systems of Display: The Making of Anatomical Knowledge in Enlightenment Britain," *British Journal for the History of Science* 46, no. 3 (2012): 360.
101. Hope, "Remarks on Botanical Lectures."

102. Berkowitz, "Systems of Display," 360.
103. October 20, 1807, Minutes of Meetings of the Faculty, 1806–1813.
104. Berkowitz, *Charles Bell,* 48.

Chapter 2. Creating a Perpetual Spring

1. Lettsom, *Grove-Hill: A Rural and Horticultural Sketch,* 9.
2. For Linnaeus's influence, see Koerner, *Linnaeus.*
3. Pettigrew, *Memoirs of the Life,* 1:167.
4. Schiebinger, *Plants and Empire,* 11.
5. Ibid.
6. Ibid.
7. Miles Ogborn, "Vegetable Empires," in Curry et al., *Worlds of Natural History,* 278.
8. Ogborn, "Vegetable Empires," 278.
9. Brockway, "Science and Colonial Expansion," 450.
10. Laidlaw, *Colonial Connections,* 31–35.
11. Arens, "Flowerbeds and Hothouses."
12. Ibid., 266.
13. Ibid.
14. Egmond, *The World of Carolus Clusius,* 212–13.
15. Maria Zytaruk, "Mary Delany: Epistolary Utterances, Cabinet Spaces, & Natural History," in Laird and Weisberg-Roberts, *Mrs. Delany and Her Circle,* 141.
16. Ibid.
17. Opitz, Bergwik, and Van Tiggelen, *Domesticity in the Making of Modern Science,* 2.
18. Llanover, *The Autobiography and Correspondence of Mary Granville, Mrs. Delany,* 2:421.
19. Alicia Weisberg-Roberts, "Introduction (1) Mrs. Delany from Source to Subject," in Laird and Weisberg-Roberts, *Mrs. Delany and Her Circle,* 1.
20. Llanover, *Autobiography and Correspondence of Mary Granville, Mrs. Delany,* 2:423.
21. Ibid., 424.
22. Ibid., 422.
23. Ibid.
24. Ibid.
25. For more on Mrs. Delany and her "mosaicks," see Laird and Weisberg-Roberts, *Mrs. Delany and Her Circle.*

26. Curtis, *Proposals for Opening by Subscription a Botanic Garden.*
27. Ibid., 10.
28. Easterby-Smith, *Cultivating Commerce.*
29. Phibbs, quoting from Sir John Parnell's *Tour of England,* 1769, in *Place-Making,* 181; Laird, "Humphry Repton at Woburn Abbey," 50.
30. Fothergill to John Bartram, October 29, 1768, in Corner and Booth, *Chain of Friendship,* 289–90.
31. Ibid., 290.
32. Pettigrew, *Memoirs of the Life,* 1:28.
33. Ibid.
34. Murphy, "Collecting Slave Traders." See also Kathleen Murphy, "A Slaving Surgeon's Collection: The Pursuit of Natural History through the British Slave Trade to Spanish America," in Craciun and Terrall, *Curious Encounters.*
35. Murphy, "Collecting Slave Traders," 639.
36. Delbourgo, "Essay Review: Gardens of Life and Death," 114.
37. Murphy, "Collecting Slave Traders," 643.
38. Tobin, *Colonizing Nature.*
39. Hunting, "Dr John Coakley Lettsom, Plant-Collector of Camberwell," 222.
40. Margaret DeLacy, "Fothergill, John (1712–1780), Physician and Naturalist," *Oxford Dictionary of National Biography* (Oxford: Oxford University Press, October 4, 2007).
41. Ibid.
42. Corner and Booth, *Chain of Friendship,* 7.
43. Ibid., 17.
44. Ibid.
45. Ibid., 409. Also, for a detailed sketch of the role, networks, and relationships relating to Peter Collinson, see Wulf, *The Brother Gardeners,* 19–33.
46. For more information, see Alan W. Armstrong, ed. *"Forget not Mee & My Garden . . .": Selected Letters 1725–1768 of Peter Collinson, F.R.S.* (Philadelphia: American Philosophical Society, 2002). See also Wulf, *The Brother Gardeners.*
47. Many of his shells and corals were left in his will to William Hunter and are now in the Hunterian Museum in Glasgow. See chapter 6 for more on this.
48. Thompson, *Memoirs of the Life and a View of the Character of the late Dr. John Fothergill,* 19.
49. Lettsom, *The Works of John Fothergill,* 18.
50. Ibid., xix.
51. Ibid., xx.
52. Ibid.

53. Ibid.
54. Ibid., xx–xxi.
55. Thompson, *Memoirs of the Life and a View of the Character of the late Dr. John Fothergill,* 37.
56. Letter from Fothergill to William Bartram, South Carolina, October 22, 1772, in Corner and Booth, *Chain of Friendship,* 393.
57. Letter from Fothergill to Lionel Chalmers, January 7, 1774, in Corner and Booth, *Chain of Friendship,* 408.
58. Ibid.
59. Letter from Fothergill to Bartram, October 22, 1772, in Corner and Booth, *Chain of Friendship,* 393.
60. Letter from Fothergill to Humphry Marshall, March 15, 1770, in Corner and Booth, *Chain of Friendship,* 320–21.
61. Wills and Fry, *Plants,* xii.
62. Llanover, *Autobiography and Correspondence of Mary Granville, Mrs. Delany,* 2:426.
63. As quoted in Rembert, "William Pitcairn MD," 219.
64. Hull, "The Influence of Herman Boerhaave," 513.
65. Anonymous student, *Lectures on Botany by John Hope, 1777–8* (Edinburgh: RBGE Archives), 6.
66. Ibid.
67. For a biography of his life, see Rembert, "William Pitcairn MD."
68. William Pitcairn, *The Royal Society,* EC/1770/08.
69. Anne Dulau, "William Hunter: A Brief Account of His Life as an Art Collector," in Black, *My Highest Pleasure,* 21.
70. Rembert, "William Pitcairn MD," 222.
71. Letter to John Hope from Dr. Pitcairn, December 22, 1777, RBGE Archives, GD253/144/12/6.
72. Ibid.
73. Ibid.
74. Easterby-Smith, *Cultivating Commerce,* 101.
75. Thompson, *Memoirs of the Life and a View of the Character of the late Dr. John Fothergill,* 37.
76. Rembert, "William Pitcairn," 221.
77. Aiton, *Hortus Kewensis;* and Desmond, *Kew,* 105–6.
78. Letter 113 in Chambers, *Scientific Correspondence of Sir Joseph Banks,* 1:135.
79. Lettsom, *The Works of John Fothergill,* xxii.
80. Letter from Fothergill to Bartram, Autumn 1772, in Corner and Booth, *Chain of Friendship,* 390.

81. Llanover, *Autobiography and Correspondence of Mary Granville, Mrs. Delany,* 2:424.
82. Letter from Fothergill to Bartram, January 13, 1770, in Corner and Booth, *Chain of Friendship,* 317.
83. Ibid.
84. Letter from Fothergill to Bartram, March 19, 1770, in Corner and Booth, *Chain of Friendship,* 321.
85. Letter from Fothergill to Bartram, Autumn 1772, in Corner and Booth, *Chain of Friendship,* 389.
86. Letter from Fothergill to Bartram, 1774, in Corner and Booth, *Chain of Friendship,* 415.
87. Ibid., 416n3.
88. Lettsom, *Grove-Hill: An Horticultural Sketch,* 24.
89. See Ben-Amos, *The Culture of Giving.*
90. Hunting, "Dr John Coakley Lettsom, Plant-Collector of Camberwell," 227.
91. Wilkinson, "William Withering (1741–1799) and Edgbaston Hall."
92. Ibid., 301; and Lee, "William Withering (1741–1799): A Birmingham Lunatic."
93. Paget, *John Hunter,* 87.
94. Caroline Grigson outlines the wide variety of animal ownership during George III's reign, in *Menagerie,* 84.
95. Ibid., 91.
96. Ibid., 103.
97. Festing, "Menageries and the Landscape Garden," 105.
98. Ibid., 108. Festing also notes that women were the driving force behind more than a third of the menageries based on her research.
99. Chambers, *Plans, Elevations, Sections, and Perspective Views,* 3.
100. Ibid., 4.
101. Desmond, *Kew,* 75. Also Grigson, *Menagerie,* 171.
102. Dobson, "John Hunter's Animals," 482.
103. Hunter, "Account of an Extraordinary Pheasant," 533.
104. With thanks to Alice Marples for pointing me toward the examples of animals in Sloane's garden. Poliquin, *Beaver,* 15.
105. Mortimer, "III. The Anatomy of a Female Beaver."
106. Ibid., 179.
107. Christopher Plumb, "'Strange and Wonderful': Encountering the Elephant in Britain, 1675–1830," *Journal for Eighteenth-Century Studies* 33 (2010): 527.
108. Ibid., 180.
109. Smith was made a fellow of the Linnean Society in 1793. See Ray Desmond, *Dictionary of British and Irish Botanists and Horticulturists: Including Plant*

Collectors, Flower Painters, and Garden Designers (London: Natural History Museum, 1994), 2757.
110. Mark Harrison, "The Calcutta Botanic Garden and the Wider World, 1817–46," in Das Gupta, *Science and Modern India,* 237.
111. Easterby-Smith, *Cultivating Commerce.*
112. Ibid., 28.
113. For an expanded discussion on gardeners and expertise in this period, see Hickman, "'The Want of a Proper Gardiner.'"
114. Keevil, "Archibald Menzies," 796.
115. Ibid.
116. Noltie, *John Hope,* 26.
117. Ibid.
118. Ibid.
119. Murphy, "Collecting Slave Traders"; and Murphy, "A Slaving Surgeon's Collection."
120. As quoted in Williams, *Naturalists at Sea,* 134.
121. Ibid., 125.
122. Ibid.
123. Ibid.
124. Dr. Lettsom to Dr. Watson, London, September 3, 1795, in Pettigrew, *Memoirs of the Life,* 3:340.
125. Lettsom, *The Works of John Fothergill,* xviii.
126. Letter from Lettsom to Rev. J. Plumtre, December 1, 1800, in Pettigrew, *Memoirs of the Life,* 2:106–7.
127. McDonagh, *Elite Women and the Agricultural Landscape.*
128. Lettsom's obituary of Fothergill in *Gentleman's Magazine* 51 (April 1781): 167.
129. Corner and Booth, *Chain of Friendship,* 20.
130. Quoted in Easterby-Smith, *Cultivating Commerce,* 44.
131. Ibid.
132. Letter from Fothergill to John Bartram, October 29, 1768, in Corner and Booth, *Chain of Friendship,* 289.
133. Ibid., 20.
134. Ibid., 289.
135. Ibid.
136. "Deaths," *Gentleman's Magazine* 51 (April 1781): 194.
137. See Hickman, "'The Want of a Proper Gardiner.'"
138. James Hack Tuke, *A Sketch of the Life of John Fothergill* (London: Harris, 1879), 32.

Chapter 3. For "Curiosity and Instruction"

1. Letter from Lettsom to Rev. J. Plumtre, December 1, 1800, in Pettigrew, *Memoirs of the Life,* 2:106–7.
2. Ibid.
3. Hunting, "Dr John Coakley Lettsom, Plant-Collector of Camberwell," 223.
4. Wulf, *Brother Gardeners,* 229.
5. Adrian Tinniswood, *The Polite Tourist;* Jacques, *Gardens of Court and Country.*
6. Jacques, *Gardens of Court and Country,* 21.
7. Seeley, *Description of the Gardens of Lord Viscount Cobham at Stow.*
8. Anonymous, *A Short Account of the Principal Seats and Gardens.*
9. Jacques, *Gardens of Court and Country,* 24.
10. Ibid.
11. The Hon. Mrs. Boscawen to Mrs. Delany, October 14, 1776, in Llanover, *Autobiography and Correspondence of Mary Granville, Mrs. Delany,* 2:264–65.
12. Jacques, *Gardens of Court and Country,* 24.
13. Tinniswood, *The Polite Tourist,* 91.
14. The introduction of ticketing has been discussed in detail by Tinniswood, ibid.
15. The Hon. Mrs. Boscawen to Mrs. Delany, October 14, 1776.
16. Thompson, *Memoirs of the Life and a View of the Character of the late Dr. John Fothergill,* 39.
17. Nichols, *Illustrations of the Literary History of the Eighteenth Century,* 2:804.
18. Classen, "Museum Manners," 906.
19. Ibid., particularly 904–5.
20. Ibid., 898.
21. Barre, "Sir Samuel Hellier."
22. Ibid., 311.
23. Ibid., 142.
24. Ibid., 311.
25. Mowl and Barre, *Historic Gardens of England,* 142.
26. Quoted in Tinniswood, *Polite Tourist,* 94.
27. Barre, "Sir Samuel Hellier," 312–13.
28. "Botanic Garden," *Caledonian Mercury,* May 4, 1782.
29. Ibid.
30. Mark Purcell, *The Country House Library* (New Haven, CT: Yale University Press, 2017), 195.
31. Ibid.
32. Ibid., 229.

33. Greig, "'All Together and All Distinct.'"
34. Anne Goldgar, "The British Museum and the Virtual Representation of Culture in the Eighteenth Century," *Albion: A Quarterly Journal Concerned with British Studies* 32, no. 2 (2000).
35. Greig, "'All Together and All Distinct,'" 74.
36. Lettsom to Sir Mordaunt Martin, March 13, 1790, in Pettigrew, *Memoirs of the Life*, 2:24.
37. Elliott, *Enlightenment, Modernity and Science*, 124.
38. Ibid.
39. Edwards, *Tabulae Distantiae . . . (Companion from London to Brighthelmston)*.
40. Ibid.
41. Ibid.
42. Lettsom, *Grove-Hill: An Horticultural Sketch*.
43. Country News, *Leeds Intelligencer*, May 20, 1766.
44. Advertisement, *Caledonian Mercury*, November 18, 1767.
45. Ibid.
46. Easterby-Smith, *Cultivating Commerce*, 6.
47. Faulkner, *An Historical and Topographical Description of Chelsea*, 29.
48. Curtis, *General Indexes to the Plants*, vi.
49. Desmond, "William Curtis (1746–1799)," 7.
50. Clark, "William Curtis's London Botanic Gardens."
51. Abraham Rees, *The Cyclopædia; or, Universal Dictionary of Arts, Sciences, and Literature*, vol. 10 (London: Longman, Hurst, Rees, Orme & Brown, 1819), 610.
52. For a detailed history of the Society of Apothecaries and their garden, see Minter, *The Apothecaries' Garden*.
53. Ibid., 15.
54. Curtis, *Proposals for Opening by Subscription a Botanic*, title page.
55. Ibid., 7.
56. Ibid.
57. Curtis, *Flora Londinensis*, preface.
58. Elliott, *Enlightenment, Modernity and Science*, 142.
59. Pelling, "Collecting the World," 102.
60. Ibid.; and Laird, *A Natural History of English Gardening*.
61. *Norfolk Chronicle*, June 19, 1779.
62. Ibid.
63. See Liverpool (1802) and Hull (1812) Botanic Gardens; more detail on both can be found in Elliott, *Enlightenment, Modernity and Science*, 153–63.

64. Elliott, Watkins, and Daniels, *The British Arboretum,* 65.
65. Ibid., 64.
66. William Salisbury to Mr. Urban, *Gentleman's Magazine,* August 1, 1810, 114.
67. For more on rational recreation and parks, see, for example, Hazel Conway, *People's Parks: The Design and Development of Victorian Parks in Britain* (Cambridge: Cambridge University Press, 1991); and Carole O'Reilly "'We Have Gone Recreation Mad': The Consumption of Leisure and Popular Entertainment in Municipal Public Parks in Early Twentieth Century Britain," *International Journal of Regional and Local History* 8, no. 2 (2013).
68. Ibid.
69. John Claudius Loudon, "Hints for a National Garden Laid before the Linnean Society," Read to the Society on December 17, 1811, Linnean Society Archives, paper no. 416.
70. Simo, *Loudon and the Landscape,* 105; Laird, "John Claudius Loudon," 248; and Elliott, *Enlightenment, Modernity and Science,* 164. The concept is further explored in relation to nineteenth-century arboretums throughout Elliott, Watkins, and Daniels, *The British Arboretum.*
71. Simo, *Loudon and the Landscape,* 4.
72. Ibid., 4.
73. For more on this, see Brent Elliott, *The Royal Horticultural Society: A History, 1804–2004* (Chichester, UK: Phillimore, 2004).
74. Ibid., 13 and 66–67.
75. Moore, *La Mortola in the Footsteps of Thomas Hanbury.*
76. Loudon, "Hints for a National Garden," 13–14. Underlining in the quotation is by Loudon's hand.
77. John Claudius Loudon, *On the Laying Out, Planting, and Managing of Cemeteries: And on the Improvement of Churchyards* (London: Longmans, Green, 1847), 12–13.
78. Elliott, "'Improvement, Always and Everywhere,'" 398.
79. Loudon, "Hints for a National Garden," 5.
80. Ibid., 10.
81. Ibid.
82. Ibid., 17.
83. Jordan, "Public Parks," 85.
84. Royal Botanic Institution of Glasgow, *Companion to the Glasgow Botanic Garden, or Popular Notices of Some of the More Remarkable Plants Contained in It* (Glasgow: Smith, ca. 1818), 12.
85. Ibid., 10.

Chapter 4. "Hints or Directions"

1. Pettigrew, *Memoirs of the Life,* 1:167.
2. In particular the work of Staffan Müller-Wille and Isabelle Charmantier on Linnaeus's use of paper for cataloging, representing, and organizing research is relevant here, including Müller-Wille and Charmantier, "Lists as Research Technologies," *Isis* 103, no. 4 (2012); Charmantier, "Carl Linnaeus and the Visual Representation of Nature," *Historical Studies in the Natural Sciences* 41 (2011); and Charmantier and Müller-Wille, "Carl Linnaeus's Botanical Paper Slips (1763–1774)," *Intellectual History Review* 24 (2014).
3. Lettsom, *Grove-Hill: An Horticultural Sketch,* iv.
4. Ibid.
5. Hunt, *Greater Perfections,* 125.
6. Ibid.
7. Ibid., 136.
8. Emma Spary, "'The "Nature' of Enlightenment," in Clark, Golinski, and Schaffer, *The Sciences in Enlightened Europe,* 295.
9. David Wallace, "Bourgeois Tragedy or Sentimental Melodrama? The Significance of George Lillo's *The London Merchant,*" *Eighteenth-Century Studies* 25, no. 2 (Winter 1991–1992).
10. Maurice, *Grove-Hill, A Descriptive Poem,* 5–6.
11. Hunting, "Dr John Coakley Lettsom, Plant-Collector of Camberwell."
12. Anonymous, "John Coakley Lettsom MD," 471.
13. Lettsom, *Hints Designed to Promote Beneficence, Temperance and Medical Science.*
14. Anderson, *Touring and Publicizing England's Country Houses,* 9.
15. Ibid., 19.
16. Dr. Lettsom to Dr. Walker, London, September 3, 1795, in Pettigrew, *Memoirs of the Life,* 3:341.
17. Lettsom to Mordaunt Martin, February 5, 1789, in Pettigrew, *Memoirs of the Life,* 2:10–11.
18. For a detailed account of the fascinating and irregular life of John Hill, see George Rousseau, *The Notorious Sir John Hill: The Man Destroyed by Ambition in the Era of Celebrity* (Lanham, MD: Lehigh University Press, 2012).
19. Roy Porter, *Quacks: Fakers and Charlatans in English Medicine* (Stroud, UK: Tempus, 2001), 180–92.
20. Ibid.
21. Anderson, *Touring and Publicizing England's Country Houses,* 62.
22. Lettsom, *Grove-Hill: A Rural and Horticultural Sketch,* 15.

23. Ibid., 9.
24. David Lambert, "The Prospect of Trade: The Merchant Gardeners of Bristol in the Second Half of the Eighteenth Century," in Michel Conan, ed., *Bourgeois and Aristocratic Cultural Encounters in Garden Art, 1550–1850* (Washington, DC: Dumbarton Oaks Research Library and Collection, 2002).
25. Lettsom, *Grove-Hill: A Rural and Horticultural Sketch,* 19.
26. Pettigrew, *Memoirs of the Life,* 1:164.
27. Dr. Lettsom to Sir M. Martin, December 5, 1790, in Pettigrew, *Memoirs of the Life,* 2:33.
28. Dr. Lettsom to Sir M. Martin, January 20, 1791, in Pettigrew, *Memoirs of the Life,* 2:35.
29. For more on this relationship, see Hunt, for example "Emblem and Expressionism in the Eighteenth-Century Landscape Garden," *Eighteenth-Century Studies* 4, no. 3 (Spring 1971).
30. Maurice, *Grove-Hill, A Descriptive Poem,* 18.
31. Hunt, *Greater Perfections,* 125.
32. Maurice, *Grove-Hill, A Descriptive Poem,* preface.
33. Lettsom, *Grove-Hill: An Horticultural Sketch,* 15.
34. Lambert, "'The Poet's Feeling.'"
35. Hunting, "Dr John Coakley Lettsom, Plant-Collector of Camberwell," 234n28.
36. Hunt, "Emblem and Expressionism in the Eighteenth-Century Landscape Garden," 310.
37. Maurice, *Grove-Hill, A Descriptive Poem,* 7.
38. Lettsom, *Grove-Hill: An Horticultural Sketch,* 5.
39. Maurice, *Grove-Hill, A Descriptive Poem,* 14.
40. For more on this, see Carole Rawcliffe, "'Delectable Sightes and Fragrant Smelles': Gardens and Health in Late Medieval and Early Modern England," *Garden History* 36, no. 1 (2008).
41. Maurice, *Grove-Hill, A Descriptive Poem,* 9.
42. Ibid., 19 and 28.
43. Lettsom, *Grove-Hill: An Horticultural Sketch,* 19.
44. Lettsom, *Grove-Hill: A Rural and Horticultural Sketch,* 22.
45. Spary, "The 'Nature' of Enlightenment," 299.
46. Van Sant, *Eighteenth-Century Sensibility and the Novel,* 5.
47. Anonymous, "Grove Hill; An Horticultural Sketch," 533.
48. Letter from Rev. Plumptre to Lettsom, November 27, 1804, in Pettigrew, *Memoirs of the Life,* 2:104.
49. Anonymous, "Grove Hill; An Horticultural Sketch," 533.
50. Lettsom, *Grove-Hill: A Rural and Horticultural Sketch,* 27.

51. Lucia Tongiori Tomasi, "Gardens of Knowledge and the République des Gens de Sciences," in Conan, *Baroque Garden Cultures,* 98.
52. Ibid.
53. For more on the catalogs of nurserymen in this period, see Easterby-Smith, *Cultivating Commerce,* 51–63.
54. Lettsom, *The Hortus Uptonensis.*
55. Lettsom, *The Works of John Fothergill,* 495.
56. Hunting, "Dr John Coakley Lettsom, Plant-Collector of Camberwell," 222.
57. Ibid.
58. Ibid.
59. Lettsom, *The Naturalist's and Traveller's Companion.*
60. This has been expertly described by Blatchly and James, "The Beeston-Coyte Hortus Botanicus Gippovicensis."
61. Defoe, *Tour through the Eastern Counties of England,* 97.
62. Blatchly and James, "The Beeston-Coyte Hortus Botanicus Gippovicensis," 339.
63. Ibid., 349.
64. Ibid.
65. Ibid.
66. Johnson, "Labels and Planting Regimes," 68.
67. Curtis, *Proposals for Opening by Subscription a Botanic Garden,* 8–9.
68. Stearn, "Sources of Information about Botanic Gardens and Herbaria," 229.
69. Curtis, *Proposals for Opening by Subscription a Botanic Garden,* 17.
70. Ibid., 14.
71. Ibid.
72. Francis Buchanan, Notes Taken from Dr. John Hope's Lectures on Botany (Summer 1780), RBGE archives.
73. Bleichmar, "Learning to Look," 90.
74. Johnson, *Nature Displaced, Nature Displayed,* 56.
75. O'Kane, "The Irish Botanical Garden," 447.
76. A detailed description of the creation of Glasnevin botanic garden and Foster's role has already been explored in Nelson and McCracken, *The Brightest Jewel;* and Johnson, *Nature Displaced, Nature Displayed.*
77. Nuala C. Johnson, "Grand Design(er)s: David Moore, Natural Theology and the Royal Botanic Gardens in Glasnevin, Dublin, 1838–1879." *Cultural Geographies* 14, no. 1 (January 2007): 34, https://doi.org/10.1177/1474474007072818.
78. O'Kane, "The Irish Botanical Garden," 446.
79. Johnston, *Nature Displaced, Nature Displayed,* 46.
80. Mark Purcell, *The Country House Library* (New Haven, CT: Yale University Press, 2017), 154–55.

81. Ibid.
82. Hunting, "Dr John Coakley Lettsom, Plant-Collector of Camberwell," 233.
83. Daston, "The Sciences of the Archive," 160.
84. Edwards, *Tabulae Distantiae . . . (Companion from London to Brighthelmston)*, 5.
85. William Noblett, "William Curtis's Botanical Library," *The Library*, s6-IX, no. 1 (1987): 3.
86. Daston, "The Sciences of the Archive," 157.
87. Elliott, *Enlightenment, Modernity and Science*, 150.
88. Noblett, "William Curtis's Botanical Library," 5.
89. Ibid.
90. Ibid.
91. John Evelyn, *Memoirs Illustrative of the Life and Writings of John Evelyn: Comprising His Diary, from the Year 1641 to 1705–6, and a Selection of his Familiar Letters . . .*, 5 vols. (London: Colburn, 1827), 1:136–37.
92. MacGregor, *Curiosity and Enlightenment*, 37.
93. Nichols, *Illustrations of the Literary History of the Eighteenth Century*, 2:665.
94. *A Catalogue of the Greater Portion of the Library*.
95. Lettsom, *Grove-Hill: An Horticultural Sketch*, 13.
96. "Extract of a Letter from Cambridge," *Jackson's Oxford Journal* 23 (January 1773). Professor Martin fulfilled a similar superintendent role to John Hope in Edinburgh and also ran lectures in the botanic garden.
97. Bleichmar, "Learning to Look," 87.
98. Thornton and Lee, *An Introduction to the Science of Botany*, xiv.
99. Coulton, *Curiosity, Commerce and Conversation*, 22.
100. Thornton and Lee, *An Introduction to the Science of Botany*, xiv.
101. Coulton, *Curiosity, Commerce and Conversation*, 3.
102. Elliott, *Enlightenment, Modernity and Science*, 151.
103. Ibid.
104. Lettsom, *The Works of John Fothergill*, xxi.
105. Henry Oakeley, Jane Knowles, Anthony Dayan, and Michael de Swiet, *A Garden of Medicinal Plants* (London: Little Brown, 2015), 40.
106. "Prospectus for the Dublin Society's Botanic Garden, Giving an Account of the Sections into Which It Will Be Divided," ca. 1795, PRONI D562/7829 C.
107. Ibid.
108. Ibid.
109. Ibid.
110. Forsythe and Cole, *Discover the Botanic Cottage*, 2.

Chapter 5. For *Dulce* and *Utile*

1. Pettigrew, *Memoirs of the Life,* 1:106.
2. Ibid.
3. Ibid.
4. Ibid.
5. Ibid.
6. Timothy Raylor, "Samuel Hartlib and the Commonwealth of Bees," in Leslie and Raylor, *Culture and Cultivation in Early Modern England,* 92.
7. Weston, *Tracts on Practical Agriculture and Gardening,* xvi.
8. Ibid., xvi–xvii.
9. Ibid., xvii.
10. See Spooner's groundbreaking study on this area, *Regions and Designed Landscapes in Georgian England,* 21.
11. Cited in Williamson, *Polite Landscapes,* 122.
12. Young and General Board of Agriculture, *General View of the Agriculture of Hertfordshire,* 233–34.
13. John Gascoigne, *Science in the Service of Empire: Joseph Banks, the British State and the Uses of Science in the Age of Revolution* (New York: Cambridge University Press, 1998), 130.
14. Drayton, *Nature's Government,* 88. For more on the king's interest in agriculture, see pp. 87–89, and for more on the political appropriation of both farming and the idea of improvement, see pp. 148–51.
15. Young and General Board of Agriculture, *General View of the Agriculture of Hertfordshire,* 234.
16. Spooner, *Regions and Designed Landscapes in Georgian England,* 18.
17. Williamson, *Polite Landscapes,* 121.
18. Bucknell, "The Mid-Eighteenth-Century Georgic and Agricultural Improvement," 335.
19. See Brown's in-depth analysis of these interrelationships, in *Performing Medicine,* particularly 48–81.
20. Spooner, *Regions and Designed Landscapes in Georgian England,* 21.
21. Young, *Annals of Agriculture and Other Useful Arts,* 1:80.
22. Monk, *General View of the Agriculture of the County of Leicester,* 60.
23. Thompson, *An Account of the Life, Lectures and Writings of William Cullen,* 1:565.
24. Withers, "William Cullen's Agricultural Lectures and Writings," 148.
25. Ibid., 156.
26. Chaplin, "John Hunter and the 'Museum Oeconomy,'" 110.

27. *A Dissertation on the Chief Obstacles to the Improvement of Land and Introducing Better Methods of Agriculture throughout Scotland,* quoted in Fussell, *More Old English Farming Books,* 35.
28. For an expert reading on design and the transition to nineteenth-century botanic gardens, see Johnson, *Nature Displaced, Nature Displayed;* and for a history of rhubarb, see Foust, *Rhubarb,* 117.
29. For more information on the fascinating history of rhubarb, see Foust, *Rhubarb.*
30. Letter from Dr. Collingwood, "table of the comparative strength of different rhubarbs" tried on his patients 1779–1785, Royal Society of Arts Archive, RSA/PR/MC/103/10/575.
31. Ibid.
32. Ibid.
33. Dr. Benjamin Rush to Lettsom, October 24, 1788, in Pettigrew, *Memoirs of the Life,* 3:188–89.
34. Ibid.
35. One of the reasons American plants were so popular in this period was that collectors believed that they could be grown outdoors in open ground, rather than in hothouses. Wulf, *The Brother Gardeners,* 27.
36. Letter from H. J. De Salis inquiring about the conditions of membership of the society and about his experiments with rhubarb, October 25, 1797, Royal Society of Arts Archive, RSA/AD/MA/100/10/348.
37. Ibid.
38. J. W. Hulke, "The Hunterian Oration of John Hunter the Biologist," *British Medical Journal,* February 23, 1895, 405.
39. Review of "*Hortus Botanicus* Gippovicensis; *or, a Systematical Enumeration,*" *Gentleman's Magazine* 66, no. 6 (June 1796): 500.
40. Ibid.
41. Blatchly and James, "The Beeston-Coyte Hortus Botanicus Gippovicensis," 350.
42. Ibid., 349.
43. Ibid.
44. Watson, "An Account of Some Experiments," 203–6.
45. Curtis, *Practical Observations on the British Grasses,* 4–5.
46. Ibid., 5.
47. Ibid.
48. Curtis, *Flora Londinensis,* preface.
49. Iain Milne, "Home, Francis (1719–1813), Physician," *Oxford Dictionary of National Biography* (Oxford: Oxford University Press, September 23, 2004).

50. See Brown, *Performing Medicine,* 51.
51. With thanks to Andrew Legg for drawing my attention to this letter from Miller, *Letters of Edward Jenner,* 4–5.
52. Ibid.
53. Ibid.
54. Ibid.
55. Humphry Davy and John Davy, *The Collected Works of Sir Humphry Davy, Bart. . . . Edited by his Brother, John Davy,* 9 vols. (London: Smith, Elder, 1839), 1:446–47.
56. Ibid.
57. Spooner, *Regions and Designed Landscapes in Georgian England,* 19.
58. Tarlow, *The Archaeology of Improvement in Britain,* 12.
59. Williamson, *The Transformation of Rural England,* 19.
60. Curtis, *Proposals for Opening by Subscription a Botanic Garden,* 4.
61. Thompson, *An Account of the Life, Lectures and Writings of William Cullen,* 1:565.
62. Lausen-Higgins, "Sylva Botanica," 230.
63. Weston, *Tracts on Practical Agriculture and Gardening,* iii.
64. Easterby-Smith and Senior, "The Cultural Production of Knowledge," 471.
65. Chambers, *The Planters of the English Landscape,* 3.
66. Ibid., 11.
67. Ibid., 8.
68. See Brown, *Performing Medicine,* particularly 55–58.
69. Letter from Rev. Dr. Thomas Lyster regarding Sir W. G. Newcomen's experiments with Chinese hempseed, March 13, 1786, Royal Society of Arts Archives, RSA/PR/MC/103/10/464.
70. Ibid.
71. A detailed description of the creation of Glasnevin botanic garden and Foster's role appears in Nelson and McCracken, *The Brightest Jewel;* and Johnson, *Nature Displaced, Nature Displayed.*
72. Baird, *General View of the Agriculture of the County of Middlesex,* 38.
73. Chaplin, "John Hunter and the 'Museum Oeconomy,'" 230.
74. Ibid.
75. Ibid., 230–31.
76. See Moore, *The Knife Man* for more detail on his experimental work.
77. Baird, *General View of the Agriculture of the County of Middlesex,* 38.
78. Ibid.
79. Chaplin, "John Hunter and the 'Museum Oeconomy,'" 230.
80. Brown, *Performing Medicine,* 51–8.

81. For a detailed account of the biology experiments Hunter conducted, see Hulke, "The Hunterian Oration."
82. Hunter, *Observations on Certain Parts of the Animal Oeconony,* 177.
83. Baird, *General View of the Agriculture of the County of Middlesex,* 39.
84. Ibid., 40.
85. Williamson, *Polite Landscapes,* 122–24.
86. Greater London Council, *Survey of London—South Kensington,* 197.
87. Moore, *The Knife Man,* 293.
88. Schupbach, "Illustrations from the Wellcome Institute Library," 351.
89. Hunter writes about his attempts to culture pearls in a letter to Joseph Banks, reprinted in Frank Buckland, *Log-Book of a Fisherman and Zoologist* (London: Chapman & Hall, 1875), 327.
90. Hunter, "Observations on Bees," 132.
91. Ibid., 132.
92. Ibid., 182–83.
93. Raylor, "Samuel Hartlib and the Commonwealth of Bees," 95.
94. Paskins, "Sentimental Industry," 92.
95. Pettigrew, *Memoirs of the Life,* 1:166.
96. O'Malley, "Introduction to John Evelyn and the 'Elysium Britannicum,'" in O'Malley and Wolschke-Bulmahn, *John Evelyn's "Elysium Britannicum,"* 28.
97. Maurice, *Grove-Hill, A Descriptive Poem,* 31.
98. Ibid., 42 and plate opposite, 30.
99. John Coakley Lettsom, *Hints for Promoting a Bee Society* (London: Darton and Harvey, 1796), 8.
100. Ibid., 9.
101. Given Lettsom's release of slaves in 1768 from the plantation he inherited in Tortola when he was twenty-three years old, we can also speculate that there may have been a further ethical reason for encouraging the local production of honey as a source of sweetness rather than sugar, although there is no evidence of this within the pamphlet.
102. Lettsom, *Hints for Promoting a Bee Society,* 9.
103. Ibid.
104. Raylor, "Samuel Hartlib and the Commonwealth of Bees," 106.
105. Maurice, *Grove-Hill, A Descriptive Poem,* 31.
106. Raylor, "Samuel Hartlib and the Commonwealth of Bees," 100.
107. Hunting, "Dr John Coakley Lettsom, Plant-Collector of Camberwell," 227.
108. Lettsom, *Of the Improvement of Medicine in London,* 19–20.
109. Ibid., 26.

Chapter 6. This "Terrestrial Elysium"

1. Anonymous, *Gentleman's Magazine* 74.
2. Oliver Pickering, "'The Quakers Tea Table Overturned': An Eighteenth-Century Moral Satire," *Quaker Studies* 17, no. 2 (2013), 252.
3. Anonymous, *Gentleman's Magazine* 74, 473.
4. Ibid.
5. Ibid., 473–74.
6. Ibid.
7. For a detailed discussion of the pleasure ground at night with its lavish parties, illuminations, and fireworks, see Felus, *The Secret Life of the Georgian Garden,* 190–213.
8. Mrs. Delany to Mrs. Dewes, Delville, February 15, 1752, in Llanover, *Autobiography and Correspondence of Mary Granville, Mrs. Delany,* 3:88–89.
9. Ibid.
10. Ibid., 89.
11. Mrs. Delany to Mrs. Dewes, Delville, February 7, 1752, in Llanover, *Autobiography and Correspondence of Mary Granville, Mrs. Delany,* 3:1–3.
12. Mrs. Delany to Mrs. Port of Ilam, June 1774, in Llanover, *Autobiography and Correspondence of Mary Granville, Mrs. Delany,* 2:1–3.
13. Ibid.
14. Ibid.
15. Thomas Byerley, ed., "Garrick's Villa, Hampton," Saturday, August 2, 1823, in *The Mirror of Literature, Amusement, and Instruction* (London: J. Limbird, 1827), 2:161.
16. Brown, *Performing Medicine,* 74.
17. Clubability in relation to medical practitioners is explored by Brown in ibid., 24–32.
18. Cooper, *The Life of Sir Astley Cooper,* 2:125–26.
19. Ibid., 128.
20. Ibid., 129.
21. Mark Silverman, "The Tradition of the Gold-headed Cane," *Pharos,* Winter 2001, 43.
22. Cooper, *The Life of Sir Astley Cooper,* 129.
23. Nichols, *Illustrations of the Literary History of the Eighteenth Century,* 2:663–64.
24. James Boswell, "Ode to Charles Dilly," *Gentleman's Magazine* 69 (April 1791): 367.
25. Nichols, *Illustrations of the Literary History of the Eighteenth Century,* 2:664.
26. Boswell, "Ode to Charles Dilly," 367.

27. Nichols, *Illustrations of the Literary History of the Eighteenth Century,* 2:664.
28. See chapter 3 for more on this.
29. Anonymous, *Gentleman's Magazine* 74, 474.
30. Ibid.
31. Brinsley Ford, "The Dartmouth Collection of Drawings by Richard Wilson," *Burlington Magazine* 90, no. 549 (1948).
32. Sweet, *Antiquaries,* 185.
33. Ibid.
34. Ibid., 47.
35. Ibid., 57.
36. Coutu, *Then and Now,* 45.
37. Daston, "The Sciences of the Archive," 162.
38. Hunting, "Dr John Coakley Lettsom, Plant-Collector of Camberwell," 228.
39. Historic England, "Kew Observatory" (first listed 1950), National Heritage List for England, no. 1357729.
40. Hunting, "Dr John Coakley Lettsom, Plant-Collector of Camberwell," 228.
41. Gillespie, "The Rise and Fall of Cork Model Collections in Britain," 129–30.
42. Davenhall, "James Ferguson: A Commemoration," 185–86.
43. The proposal states, "We whose names are underwritten, being Members of the Royal Society, do on our personal knowledge recommend Dr John Coakley Lettsom, of Great Eastcheap, as a Gentleman well qualified to make a useful Member of this Society. Proposers: B Franklin; Wm Watson Junr; John Ellis; J Colebrooke; Dan Solander; J Ives; Mar Tunstall; James Ferguson." Held at the Royal Society, GB 117, EC/1773/28.
44. Lettsom, "Anecdotes of Thomas Jefferson," 178–80.
45. Ibid., 180.
46. Pettigrew, *Memoirs of the Life,* 1:121–22.
47. Ibid., 2:122–23.
48. All quotations in this section are from Lettsom's Diary, January 1812–December 1813, Wellcome Library AMS/MF/4/17.
49. Harris, *Weatherland,* 165–70 and 208–9.
50. These and other experiments are discussed in Lloyd Allan Wells, "'Why Not Try the Experiment?' The Scientific Education of Edward Jenner," *Proceedings of the American Philosophical Society* 118, no. 2 (1974).
51. James F. Palmer, *The Works of John Hunter,* 4 vols. (London: Longman, Rees, Brown, Orme . . . , 1835), 1:70.
52. Lorraine Daston, "The Empire of Observation, 1600–1800," Lorraine Daston and Elizabeth Lunbeck, eds., *Histories of Scientific Observation* (Chicago: University of Chicago Press, 2011), 101.
53. Baron, *The Life of Edward Jenner,* 1:18.

54. With thanks to Dr. Jenner's house staff for bringing this to my attention; the letter is held in the New York Academy of Medicine, MS. 1277.
55. Ibid.
56. For more on this, see Hickman, "The Garden as a Laboratory."
57. Wright, *Universal Architecture: Six Original Designs of Arbours.*
58. Jackson-Stops, *An English Arcadia,* 77–78.
59. Ibid.
60. Brown, *Performing Medicine,* 226.
61. Ibid.
62. Joseph Houlton, "Proposal to Establish a Garden of Medical Botany in London," *The Lancet* 12, no. 314 (1829): 721.
63. Ibid.
64. William Howison, "Remarks on the Advantage to Medical Men of Possessing a Knowledge of Botany," *The Lancet* 28, no. 734 (1837): 936.
65. Ibid.
66. Ibid.
67. Jonathan Reinarz, "The Age of Museum Medicine: The Rise and Fall of the Medical Museum at Birmingham's School of Medicine," *Social History of Medicine* 18, no. 3 (2005): 419–37.
68. G. B. Knowles, "Observations on the Importance of the Study of Botany, as a Branch of Medical Education; Addressed to the Botanical Class in Queen's College, Birmingham, at the close of the late Summer Session, 1845," *The Lancet* 46, no. 1145 (1845): 144–45.
69. F. Wm. Headland, "On Medical Education," *The Lancet* 95, no. 2426 (1870): 300.
70. Historic Environment Scotland, "Inventory of Gardens and Designed Landscapes," "Glasgow Botanic Gardens" (1987), GDL00190.
71. Royal Botanic Institution of Glasgow, *Companion to the Glasgow Botanic Garden, or Popular Notices of Some of the More Remarkable Plants Contained in It* (Glasgow: Smith, c.1818), 5.
72. Hunting, "Dr John Coakley Lettsom, Plant-Collector of Camberwell," 233.
73. Nichols, *Illustrations of the Literary History of the Eighteenth Century,* 665.
74. Ibid.
75. A. P. Baggs, Diane K. Bolton, and Patricia E. C. Croot, "Islington: Growth," in T. F. T. Baker and C. R. Elrington, eds., *A History of the County of Middlesex: Volume 8, Islington and Stoke Newington Parishes* (London: Victoria County History, 1985): *British History Online,* https://www.british-history.ac.uk/vch/middx/vol8/, 9–19, accessed September 8, 2019.
76. Peter Black, "Taste and the Anatomist," in Black, *My Highest Pleasure,* 75.

Epilogue

1. Harwood et al., "Whither Garden History?"
2. Williamson in ibid., 97.
3. Gregory, Spooner, and Williamson, *Lancelot "Capability" Brown,* 38.
4. Wylie, *Landscape,* 5.
5. Key examples include Hannah Macpherson, "Walkers with Visual-Impairments in the British Countryside: Picturesque Legacies, Collective Enjoyments and Well-Being Benefits," *Journal of Rural Studies* 51 (2017); Sarah Bell, "Sensing Nature: Unravelling Metanarratives of Nature and Blindness," in Atkinson and Hunt, *Geohumanities and Health;* and Bell, "Experiencing Nature with Sight Impairment."
6. Ruggles, *Sound and Scent in the Garden.*
7. *Environment and History—Leaping the Fence: Gardens, Parks and Environmental History,* special issue, 24, no. 1 (2018). Recent key works developing this approach include James Beattie, *Empire and Environmental Anxiety: Health, Science, Arts and Conservation in South Asia and Australasia 1800–1920* (Basingstoke, UK: Palgrave Macmillan, 2011).
8. Hannah Atkinson et al., *Race, Ethnicity & Equality in UK History: A Report and Resource for Change* (London: Royal Historical Society, October 2018).
9. Smit and Kendle, "The Eden Project."
10. The opportunities and limitations of such projects more broadly have been discussed in Kerryn Husk et al., "What Approaches to Social Prescribing Work, for Whom, and in What Circumstances? A Realist Review," *Health and Social Care in the Community* 28 (2020).
11. Historic England, *Easy Access to Historic Landscapes.*
12. Harris, *Weatherland;* Laird, *A Natural History of English Gardening.*
13. Gregory, Spooner, and Williamson, *Lancelot "Capability" Brown,* 39.
14. More on the project led by Corinne Fowler can be found at her blog *Colonial Countryside: Youth-led History,* supported by Arts Council England, https://colonialcountryside.wordpress.com/, accessed May 1, 2020.
15. Sally-Anne Huxtable, Corinne Fowler, Christo Kefalas, and Emma Slocombe, *Interim Report on the Connections between Colonialism and Properties Now in the Care of the National Trust, Including Links with Historic Slavery* (Swindon: National Trust, September 2020).
16. Tobin, *Colonizing Nature.*
17. Alexandre Antonelli, "It Is Time to Decolonise Botanical Collections," Royal Botanic Gardens, Kew, June 25, 2020, https://www.kew.org/read-and-watch/time-to-decolonise-botanical-collections, accessed November 14, 2020.

18. Jon Favreau interviewing Barack Obama, series 1, episode 15, *The Wilderness,* September 24, 2018, part of a series of podcasts produced by Crooked Media, https://crooked.com/podcast/chapter-15-the-story/, accessed May 1, 2020.
19. The *Storying the Past* blog is available at https://storyingthepast.wordpress.com/creative-histories-events/, accessed May 1, 2020.
20. Steve Poole, "Ghosts in the Garden: Locative Gameplay and Historical Interpretation from Below," *International Journal of Heritage Studies* 24, no. 3 (2017): 9.
21. Catherine Horwood, *Gardening Women: Their Stories from 1600 to the Present* (London: Virago, 2010); Madeleine Pelling, "Collecting the World" and Women's History Network, *Women's History,* special issue: *Gardening* 2, no. 13 (2019).
22. O'Reilly, *The Greening of the City,* 89.
23. Poole, "Ghosts in the Garden," 1–2.
24. The UCL database Legacies of British Slave-ownership, https://www.ucl.ac.uk/lbs/project/links/, accessed May 1, 2020; and Dresser and Hann, *Slavery and the English Country House,* which emerged from a symposium on "Slavery and the British Country House: Mapping the Current Research" held at London School of Economics in November 2009.
25. See the *Executive Summary of the Evaluation Report on Capability Brown Festival 2016* and other legacy information, http://www.capabilitybrown.org/legacy-0, accessed May 1, 2020; and an account of the Repton celebrations as a successful way of encouraging multicultural engagements with historic landscapes was published by Linden Groves, "'To Cheer the Hearts and Delight the Eyes of All,'" *Historic Gardens Review* 40 (Winter 2019/2020).
26. There is currently only one dedicated garden history postgraduate MA course run via the School of Advanced Study in London, although there are a greater number of continuing education courses and David Marsh's Grapevine Network, which runs short courses across the United Kingdom on a variety of garden history-related subjects. Concerns around the decline in academic work on the subject were raised in 2007 by Harwood et al. in "Whither Garden History?"

Selected Bibliography

Key Archives Consulted

National Botanic Gardens of Ireland, Dublin
University of Glasgow
Linnean Society
University of Oxford Herbaria, Department of Plant Sciences
Royal Botanic Garden Edinburgh
Royal Botanic Gardens, Kew
Royal Society
Royal Society of Arts
Wellcome Collection

Printed Primary Sources

A Catalogue of the Greater Portion of the Library of John Coakley Lettsom: Removed from His Residence at Camberwell Which Will Be Sold by Auction, by Leigh and S. Sotheby . . . on Monday, March 26, 1811, at Twelve o'clock, and Seven Following Days (Sunday Excepted), 1811.

Aiton, William. *Hortus Kewensis, or, A Catalogue of the Plants Cultivated in the Royal Botanic Garden at Kew.* London: Printed for George Nicol, 1789.

Anonymous. *A Short Account of the Principal Seats and Gardens in and about Richmond and Kew.* Brentford, UK: Printed and sold by P. Norbury and George Bickham, 1763.

———. *Gentleman's Magazine* 74 (Wednesday, May 23, 1804): 473–74.

———. "Grove Hill; An Horticultural Sketch. London 1794," in *The British Critic, A New Review.* London: Printed for F. and C. Rivington, 1795, 532–34.

———. "John Coakley Lettsom MD," *Gentleman's Magazine,* November 1815, 469–73.

Arnot, Hugo. *History of Edinburgh.* Edinburgh: W. Creech; London: J. Murray, 1789.

Baird, Thomas. *General View of the Agriculture of the County of Middlesex, with Observations on the Means of Its Improvement. Drawn Up for the Consideration of the Board of Agriculture and Internal Improvement.* London: J. Nichols, 1793.

Chambers, William. *Plans, Elevations, Sections, and Perspective Views of the Gardens and Buildings at Kew in Surry; The Seat of Her Royal Highness The Princess Dowager of Wales.* London: Printed by J. Haberkorn, in Grafton Street, 1763.

Cullen, William. *Lectures on the Materia Medica as Given by William Cullen.* Dublin: Printed by W. and H. Whitestone, 1761.

Curtis, Samuel. *General Indexes to the Plants Contained in the First Fifty-Three Volumes (Or Old Series Complete) of the Botanical Magazine.* London: Printed by Couchman for S. Curtis, 1828.

Curtis, William. *Flora Londinensis: Or Plates and Descriptions of Such Plants As Grow Wild in the Environs of London: With Their Places of Growth, and Times of Flowering; Their Several Names according to Linnaeus and Other Authors: With a Particular Description of Each Plant in Latin and English. To Which Are Added, Their Several Uses in Medicine, Agriculture, Rural Oeconomy, and Other Arts.* London: Printed for the Author, 1777.

———. *Practical Observations on the British Grasses, Especially Such as Are Best Adapted to the Laying Down or Improving of Meadows and Pastures: Likewise, an Enumeration of the British Grasses.* 4th ed. London: Printed for H. D. Symonds and Curtis, 1805.

———. *Proposals for Opening by Subscription a Botanic Garden to be Called the London Botanic Garden.* London: J. Andrews for the Author, 1778.

Defoe, Daniel. *Tour through the Eastern Counties of England, 1722.* London: Cassell, 1888.

Edwards, James. *Tabulae Distantiae; or, Two Tables of Lineal Distances (Description of Southwark, Lambeth, Newington, &c, Companion from London to Brighthelmston).* Dorking, UK: 1789.

Faulkner, Thomas. *An Historical and Topographical Description of Chelsea, and its Environs; Interspersed with Biographical Anecdotes of Illustrious and Eminent Persons Who Have Resided in Chelsea During the Three Preceding Centuries.* London: Printed by J. Tilling, Chelsea, 1810.

Hill, John. *An Idea of a Botanical Garden, in England: With Lectures on the Science, Without Expence to the Public, or to the Students.* London: Printed for R. Baldwin, at the Rose in Paternoster Row, 1758.

———. *Hortus Kewensis, Sistens Herbas Exoticas, Indigenasque Rariores, in Area Botanica Hortorum Augustissimae Principissae Cambriae Dotissae Apud Kew in Comitatu Surreiano Cultas.* London: Ricardum Baldwin, 1768.

Hunter, John. "Account of an Extraordinary Pheasant." *Philosophical Transactions of the Royal Society.* 70 (1780): 527–35.

———. "Observations on Bees." *Philosophical Transactions.* 82 (1792): 128–95.
———. *Observations on Certain Parts of the Animal Oeconomy.* London: Published by the Author, 1786.
Johnson, J. *Guide for Gentlemen Studying Medicine at the University of Edinburgh.* London: Printed for G. G. J. and J. Robinson; Bell and Bradfute; Mudie; Hill, Edinburgh; and W. Jones, Dublin, 1792.
Lettsom, John Coakley. "Anecdotes of Thomas Jefferson, President of the United States of America, and of The Little Turtle, Chief of the Miamis Indians; With an Account of Vaccination Among Them." *European Magazine, and London Review* 42 (1802): 178–80.
———. *Grove-Hill: An Horticultural Sketch.* London: Printed for the Author, 1794.
———. *Grove-Hill: A Rural and Horticultural Sketch.* London: Printed for Stephen Couchman, 1804.
———. *Hints Designed to Promote Beneficence, Temperance and Medical Science.* London: J. Manman, 1801.
———. *Of the Improvement of Medicine in London, on the Basis of Public Good.* London: James Phillips for E. & C. Dilly, 1775.
———. *The Hortus Uptonensis, or a Catalogue of Stove and Green-house Plants.* London, 1783.
———. *The Naturalist's and Traveller's Companion, Containing Instructions for Collecting & Preserving Objects of Natural History and for Promoting Inquiries after Human Knowledge in General.* London: E. & C. Dilly, 1772.
———. *The Works of John Fothergill . . . With Some Account of His Life.* London: Charles Dilly, 1784.
Maurice, Thomas. *Grove-Hill, A Descriptive Poem.* London: J. & A. Arch, 1799.
Monk, John. *General View of the Agriculture of the County of Leicester with Observations on the Means of its Improvement.* London: Printed by J. Nichols, 1794.
Mortimer, Cromwell. "III. The Anatomy of a Female Beaver, and an Account of Castor Found in Her," *Philosophical Transactions* 38 (1733–1734): 172–83.
Nichols, John. *Illustrations of the Literary History of the Eighteenth Century, Consisting of Authentic Memoir and Original Letters of Eminent Persons, and Intended as a Sequel to the "Literary Anecdotes."* Vol. 2. London: Nichols, Son, and Bentley, 1817.
Seeley, Benton. *Description of the Gardens of Lord Viscount Cobham at Stow.* Northampton, UK: W. Dicey, 1744.
Thompson, Gilbert. *Memoirs of the Life and a View of the Character of the Late Dr. John Fothergill.* London: Printed for T. Cadell, 1782.
Thompson, John. *An Account of the Life, Lectures and Writings of William Cullen.* Edinburgh: William Blackwood, 1832.

Thornton, Robert John, ed., and James Lee. *An Introduction to the Science of Botany Chiefly Extracted from the Works of Linnaeus; To Which Are Added, Several New Tables and Notes, and a Life of the Author [by R. J. Thornton].* 4th ed. London: Wilkie & Rivington, 1810.

Watson, William. "An Account of Some Experiments, by Mr. Miller of Cambridge, on The Sowing of Wheat," *Transactions of the Royal Society* 58 (December 31, 1768): 203–6.

Weston, Richard. *Tracts on Practical Agriculture and Gardening Particularly Addressed to Gentlemen Farmers in Great Britain with Several Useful Improvements in Stoves and Greenhouses.* London: Printed for S. Hooper, 1773.

Wright, Thomas. *Universal Architecture: Six Original Designs of Arbours.* London: Printed for the Author, 1755.

Young, Arthur. *Annals of Agriculture and Other Useful Arts.* 46 vols. London: Printed by H. Goldney, 1794.

Young, Arthur, and General Board of Agriculture. *General View of the Agriculture of Hertfordshire: Drawn Up for the Consideration of the Board of Agriculture and Internal Improvement.* London: G. and W. Nicol, 1804.

Secondary Literature

Anderson, Jocelyn. *Touring and Publicizing England's Country Houses in the Long Eighteenth Century.* New York: Bloomsbury, 2018.

Arens, Esther. "Flowerbeds and Hothouses: Botany, Gardens, and the Circulation of Knowledge in Things." *Historical Social Research* 40, no. 1 (2015): 265–83.

Atkinson, Sarah, and Rachel Hunt, eds. *Geohumanities and Health.* Cham, Switzerland: Springer Global Perspectives on Health Geography, 2019.

Baron, John. *The Life of Edward Jenner.* 2 vols. London: Henry Colburn, 1838.

Barre, Dianne. "Sir Samuel Hellier (1736–84) and His Garden Buildings: Part of The Midlands 'Garden Circuit' in the 1760s–70s?," *Garden History* 36, no. 2 (Winter 2008): 310–27.

Beattie, James, and Karen Jones. "Editorial." *Environment and History—Leaping the Fence: Gardens, Parks and Environmental History,* special issue, 24, no. 1 (2018): 1–4.

Bell, Sarag. "Experiencing Nature with Sight Impairment: Seeking Freedom from Ableism." *Environment and Planning E: Nature and Space* 2, no. 2 (2019): 304–22.

Ben-Amos, Ilana Krausman. *The Culture of Giving: Informal Support and Gift-Exchange in Early Modern England.* Cambridge: Cambridge University Press, 2008.

Berkowitz, Carin. *Charles Bell and the Anatomy of Reform*. Chicago: University of Chicago Press, 2015.

———. "Systems of Display: The Making of Anatomical Knowledge in Enlightenment Britain." *British Journal for the History of Science* 46, no. 3 (2012): 359–87.

Black, Peter, ed. *My Highest Pleasure: William Hunter's Art Collection*. Glasgow: The Hunterian, University of Glasgow, in association with Paul Holberton Publishing, 2007.

Blatchly, John, and Jenny James. "The Beeston-Coyte Hortus Botanicus Gippovicensis and Its Printed Catalogue." *Proceedings of the Suffolk Institute of Archaeology and History* 39, no. 3 (1999): 339–52.

Bleichmar, Daniela. "Learning to Look: Visual Expertise across Art and Science in Eighteenth-Century France." *Eighteenth-Century Studies* 46, no. 1 (2012): 85–111.

Boney, A. D. *The Lost Gardens of Glasgow University*. Kent: Christopher Helm, 1988.

Brockway, Lucile H. "Science and Colonial Expansion: The Role of the British Royal Botanic Gardens." *American Ethnologist*. 6, no. 3 (1979): 449–65.

Brown, Michael. *Performing Medicine: Medical Culture and Identity in Provincial England, c. 1760–1850*. Manchester, UK: Manchester University Press, 2011.

Bucknell, Clare. "The Mid-Eighteenth-Century Georgic and Agricultural Improvement." *Journal for Eighteenth-Century Studies* 36 (2013): 335–52.

Chambers, Douglas. *The Planters of the English Landscape Garden*. New Haven, CT, and London: Yale University Press, 1993.

Chambers, Neil, ed. *Scientific Correspondence of Sir Joseph Banks, 1795–1820*. 6 vols. London: Pickering and Chatto, 2006.

Chaplin, Simon. "John Hunter and the 'Museum Oeconomy,' 1750–1800." Unpublished PhD thesis, King's College London, 2009.

Clark, Kath. "William Curtis's London Botanic Gardens and *Flora Londinensis*." *London Gardener* 15 (2009–2010): 26–34.

Clark, William, Jan Golinski, and Simon Schaffer, eds. *The Sciences in Enlightened Europe*. Chicago: University of Chicago Press, 1999.

Classen, Constance. "Museum Manners: The Sensory Life of the Early Museum." *Journal of Social History* 40, no. 4 (2007): 895–914.

Conan, Michel, ed. *Baroque Garden Cultures: Emulation, Sublimation, Subversion*. Washington, DC: Dumbarton Oaks, 2005.

Cook, Alexandra. *Jean-Jacques Rousseau and Botany: The Salutary Science*. Oxford: Voltaire Foundation, 2012.

Cooper, Bransby Blake. *The Life of Sir Astley Cooper, Bart., Interspersed with*

Sketches from his Note-Books of Distinguished Contemporary Characters. 2 vols. London: John W. Parker, 1843.

Corner, Betsy, and Christopher Booth. *Chain of Friendship: Selected Letters of Dr. John Fothergill, 1735–1780.* Cambridge, MA: Harvard University Press, 1971.

Coulton, Richard. "Curiosity, Commerce and Conversation in the Writing of London Horticulturalists during the Early-Eighteenth Century." Unpublished PhD thesis, Queen Mary University of London, 2005.

Coutu, Jean. *Then and Now: Collecting and Classicism in Eighteenth-Century England.* Montreal: McGill-Queen's University Press, 2015.

Craciun, Adriana, and Mary Terrall, eds. *Curious Encounters: Voyaging, Collecting, and Making Knowledge in the Long Eighteenth Century.* Toronto: University of Toronto Press, 2019.

Curry, H., N. Jardine, J. Secord, and E. Spary, eds. *Worlds of Natural History.* Cambridge: Cambridge University Press, 2018.

Das Gupta, Uma, ed. *Science and Modern India: An Institutional History, c. 1784–1947.* Delhi: Pearson Longman, 2011.

Daston, Lorraine. "The Sciences of the Archive." *Osiris* 27, no. 1, Clio Meets Science: The Challenges of History (2012): 156–87.

Davenhall, Clive. "James Ferguson: A Commemoration." *Journal of Astronomical History and Heritage* 13, no. 3 (2010): 179–86.

Delbourgo, James. "Essay Review: Gardens of Life and Death." *British Journal for the History of Science* 43, no. 1 (2010): 113–18.

Desmond, Ray. *Kew: The History of the Royal Botanic Gardens.* London: Harvill Press, with the Royal Botanic Gardens, Kew, 1995.

———. "William Curtis (1746–1799)." *Curtis's Botanical Magazine* 4, no. 1 (1987): 7–14.

Dick, Malcolm, and Elaine Mitchell, eds. *Gardens and Green Spaces in the West Midlands since 1700.* Hatfield, UK: University of Hertfordshire Press, 2018.

Dobson, Jessie. "John Hunter's Animals." *Journal of the History of Medicine and Allied Sciences* 17, no. 4 (1962): 479–86.

Drayton, Richard. *Nature's Government: Science, Imperial Britain and the "Improvement" of the World.* New Haven, CT: Yale University Press, 2000.

Dresser, Madge, and Andrew Hann. *Slavery and the English Country House.* Swindon, UK: English Heritage, 2013.

Dugan, Holly. *The Ephemeral History of Perfume: Scent and Sense in Early Modern England.* Baltimore: Johns Hopkins University Press, 2011.

Easterby-Smith, Sarah. *Cultivating Commerce: Cultures of Botany in Britain and France, 1760–1815.* Cambridge: Cambridge University Press, 2017.

———. "Selling Beautiful Knowledge: Amateurship, Botany and the Market-Place in Eighteenth-Century France." *Journal for Eighteenth-Century Studies* 36, no. 4 (2013): 531–43.

Easterby-Smith, Sarah, and Emily Senior. "The Cultural Production of Knowledge: Contexts, Terms, Themes." *Journal for Eighteenth-Century Studies* 36, no. 4 (2013): 471–543.

Egmond, Florike. *The World of Carolus Clusius: Natural History in the Making, 1550–1610.* London: Pickering & Chatto, 2010.

Elliott, Paul. *Enlightenment, Modernity and Science: Geographies of Scientific Culture and Improvement in Georgian England.* London and New York: I. B. Tauris, 2010.

———. "'Improvement, Always and Everywhere': William George Spencer (1790–1866) and Mathematical, Geographical and Scientific Education in Nineteenth-Century England." *History of Education* 33, no. 4 (2004): 391–417.

Elliott, Paul, Charles Watkins, and Stephen Daniels. *The British Arboretum: Trees, Science and Culture in the Nineteenth Century.* London: Routledge, 2015.

———. "'Combining Science with Recreation and Pleasure': Cultural Geographies of Nineteenth-Century Arboretums." *Garden History* 35, no. 2 (2007): 6–27.

Felus, Kate. *The Secret Life of the Georgian Garden.* London: I. B. Tauris, 2016.

Festing, Sally. "Menageries and the Landscape Garden." *Journal of Garden History* 8, no. 4 (1988): 104–17.

Findlen, Paula. *Possessing Nature: Museums, Collecting, and Scientific Culture in Early Modern Italy.* Berkeley and London: University of California Press, 1994.

Fischer, Hubertus, Volker R. Remmert, and Joachim Wolschke-Bulmahn, eds. *Gardens, Knowledge and the Sciences in the Early Modern Period.* Basel, Switzerland: Birkhäuser, 2016.

Forsythe, Sutherland, and Donna Cole. *Discover the Botanic Cottage.* Edinburgh: Royal Botanic Garden Edinburgh, 2016.

Foust, Clifford. *Rhubarb: The Wondrous Drug.* Princeton, NJ: Princeton University Press, 1992.

Fussell, George Edwin. *More Old English Farming Books, 1731–1793.* London: C. Lockwood, 1950.

Gascoigne, John. *Joseph Banks and the English Enlightenment: Useful Knowledge and Polite Culture.* Cambridge: Cambridge University Press, 2003.

Gillespie, Richard. "The Rise and Fall of Cork Model Collections in Britain." *Architectural History* 60 (2017): 117–46.

Greater London Council. *Survey of London—South Kensington: Kensington Square to Earl's Court.* London: Athlone, 1986.

Gregory, Jon, Sarah Spooner, and Tom Williamson. *Lancelot "Capability" Brown: A Research Impact Review Prepared for English Heritage by The Landscape Group, University of East Anglia.* Research Report Series No. 50. Swindon, UK: Historic England, 2013.

Greig, Hannah. "'All Together and All Distinct': Public Sociability and Social Exclusivity in London's Pleasure Gardens, ca. 1740–1800." *Journal of British Studies* 51, no. 1 (2012): 50–75.

Grigson, Caroline. *Menagerie: The History of Exotic Animals in England, 1100–1837.* Oxford: Oxford University Press, 2016.

Harris, Alexandra. *Weatherland: Writers & Artists under English Skies.* London: Thames and Hudson, 2015.

Harwood, Edward, Tom Williamson, Michael Leslie, and John Dixon Hunt. "Whither Garden History?" *Studies in the History of Gardens & Designed Landscapes* 27, no. 2 (2007): 91–112.

Hickman, Clare. "Curiosity and Instruction: British and Irish Botanic Gardens and Their Audiences, 1760–1800." *Environment and History* 24, no. 1 (2018): 59–80.

———. "The Garden as a Laboratory: The Role of Domestic Gardens as Places of Scientific Exploration." *Post-Medieval Archaeology* 48, no. 1 (2014): 229–47.

———. *Therapeutic Landscapes: A History of English Hospital Gardens since 1800.* Manchester, UK: Manchester University Press, 2013.

———. "'The Want of a Proper Gardiner': Late Georgian Scottish Botanic Gardeners as Intermediaries of Medical and Scientific Knowledge." *British Journal for the History of Science* 52, no. 4 (2019): 543–67.

Historic England. *Easy Access to Historic Landscapes.* Swindon, UK: Historic England, 2015.

Horwood, Catherine. *Gardening Women: Their Stories from 1600 to the Present.* London: Virago, 2010.

Hull, Gillian. "The Influence of Herman Boerhaave." *Journal of The Royal Society of Medicine* 90, no. 9 (1997): 512–14.

Hunt, John Dixon. "Emblem and Expressionism in the Eighteenth-Century Landscape Garden." *Eighteenth-Century Studies* 4, no. 3 (1971): 294–317.

———, ed. *Garden History: Issues, Approaches, Methods.* Washington, DC: Dumbarton Oaks, 1992.

———. *Greater Perfections: The Practice of Garden Theory.* Philadelphia: University of Pennsylvania Press, 1999.

———, ed. *A World of Gardens.* London: Reaktion, 2012.

Hunting, Penelope. "Dr John Coakley Lettsom, Plant-Collector of Camberwell." *Garden History* 34, no. 2 (2006): 221–35.

Jackson-Stops, Gervase. *An English Arcadia, 1600–1990.* London: National Trust 1992.

Jacques, David. *Gardens of Court and Country: English Design, 1630–1730.* New Haven, CT: Yale University Press, 2017.

Johnson, Nuala. "Labels and Planting Regimes: Regulating Trees at Glasnevin Botanic Gardens, Dublin, 1795-1850." Supplement: Cultural and Historical Geographies of the Arboretum, *Garden History* 35 (2007): 53–70.

———. *Nature Displaced, Nature Displayed: Order and Beauty in Botanical Gardens.* London: I. B. Tauris, 2011.

Jordan, Harriet. "Public Parks, 1885-1914." *Garden History* 22, no. 1 (1994): 85–113.

Keevil, J. J. "Archibald Menzies, 1754-1842." *Bulletin of History of Medicine* 22, no. 6 (1948): 796–811.

Koerner, Lisbet. *Linnaeus: Nature and Nation.* Cambridge, MA: Harvard University Press, 1999.

Kohler, Robert. *Landscapes and Labscapes: Exploring the Lab-Field Border in Biology.* Chicago: University of Chicago Press, 2002.

Kusukawa, Sachiko. *Picturing the Book of Nature: Image, Text, and Argument in Sixteenth-Century Human Anatomy and Medical Botany.* Chicago: University of Chicago Press, 2012.

Laidlaw, Zoë. *Colonial Connections, 1815–45: Patronage, the Information Revolution and Colonial Government.* Manchester, UK: Manchester University Press, 2005.

Laird, Mark. *A Natural History of English Gardening, 1650–1800.* New Haven, CT, and London: Paul Mellon Centre for Studies in British Art, Yale University Press, 2015.

———. *Flowering of the English Landscape Garden: English Pleasure Grounds, 1720–1800.* Philadelphia: Penn Studies in Landscape Architecture, University of Pennsylvania Press, 1999.

———. "Humphry Repton at Woburn Abbey, Bedfordshire: Before and after the Red Book." *Garden History* 47, no. 1 (2019): 39–55.

———. "John Claudius Loudon (1783–1843) and the Field's Identity." *Studies in the History of Gardens & Designed Landscapes* 34, no. 3 (2014): 248–53.

Laird, Mark, and Alicia Weisberg-Roberts, eds. *Mrs. Delany and Her Circle.* New Haven, CT: Yale University Press, 2009.

Lambert, David. "'The Poet's Feeling': Aspects of the Picturesque in Contemporary Literature, 1794–1816." *Garden History* 24, no. 1 (1996): 82–99.

Lausen-Higgins, Johanna. "Sylva Botanica: Evaluation of the Lost Eighteenth-Century Leith Walk Botanic Garden Edinburgh." *Garden History* 43, no. 2 (2015): 218–36.

Lee, M. R. "William Withering (1741–1799): A Birmingham Lunatic." *Proceedings Royal College Physicians Edinburgh* 31 (2001): 77–83.

Leslie, Michael, and Timothy Raylor, eds. *Culture and Cultivation in Early Modern England.* Leicester, UK: Leicester University Press, 1992.

Livingstone, David. "Keeping Knowledge in Site." *History of Education* 39, no. 6 (2010): 779–85.

———. *Putting Science in Its Place.* Chicago: University of Chicago Press, 2003.

Livingstone, David, and Charles Withers, eds. *Geographies of Nineteenth-Century Science.* Chicago: University of Chicago Press, 2011.

Llanover, Lady Augusta, ed. *The Autobiography and Correspondence of Mary Granville, Mrs. Delany: With Interesting Reminiscences of King George the Third and Queen Charlotte.* 3 vols., 2nd series, London: Bentley, 1862.

MacGregor, Arthur. *Curiosity and Enlightenment: Collectors and Collections from the Sixteenth to the Nineteenth Century.* New Haven, CT: Yale University Press, 2007.

McDonagh, Briony. *Elite Women and the Agricultural Landscape, 1700–1830.* London: Routledge, 2016.

Miller, Genevieve, ed. *Letters of Edward Jenner and Other Documents Concerning the Early History of Vaccination.* Baltimore, MD: Johns Hopkins University Press, 1983.

Minter, Sue. *The Apothecaries' Garden: A New History of Chelsea Physic Garden.* Stroud, UK: Sutton, 2000.

Moore, Alasdair. *La Mortola in the Footsteps of Thomas Hanbury.* London: Cadogan Guides, 2004.

Moore, Wendy. *The Knife Man.* London: Bantam, 2005.

Mowl, Timothy, and Dianne Barre. *Historic Gardens of England: Staffordshire.* Bristol, UK: Redcliffe Press, 2009.

Murphy, Kathleen S. "Collecting Slave Traders: James Petiver, Natural History, and the British Slave Trade." *William and Mary Quarterly* 70, no. 4 (2013): 637–70.

Naylor, Simon. *Regionalizing Science: Placing Knowledges in Victorian England.* London: Pickering and Chatto, 2010.

Nelson, Charles, and Eileen McCracken. *The Brightest Jewel: A History of the National Botanic Gardens, Glasnevin, Dublin.* Kilkenny, Ireland: Boethius, 1987.

Noltie, Henry. *John Hope (1725–1786): Alan G. Morton's Memoir of a Scottish Botanist.* Edinburgh: Royal Botanic Garden Edinburgh, 2011.

O'Kane, Finola. "The Irish Botanical Garden: For Ireland or for Empire?" *Studies in the History of Gardens & Designed Landscape* 28, nos. 3–4 (2008): 446–55.

O'Malley, Therese, and Joachim Wolschke-Bulmahn, eds. *John Evelyn's "Elysium Britannicum" and European Gardening.* Washington, DC: Dumbarton Oaks, 1998.

Opitz, Donald L., Staffan Bergwik, and Brigitte Van Tiggelen, eds. *Domesticity in the Making of Modern Science.* Basingstoke, UK: Palgrave Macmillan, 2015.

O'Reilly, Carole. *The Greening of the City: Urban Parks and Public Leisure, 1840–1940.* New York: Routledge, 2019.

Paget, Stephen. *John Hunter: Man of Science and Surgeon, 1728–1793.* London: T. Fisher Unwin, 1897.

Paskins, Matthew. "Sentimental Industry: The Society of Arts and the Encouragement of Public Useful Knowledge, 1754–1848." Unpublished PhD thesis, University College London, 2014.

Pelling, Madeleine. "Collecting the World: Female Friendship and Domestic Craft at Bulstrode Park." *Journal for Eighteenth-Century Studies* 41, no. 1 (2018): 101–20.

Pettigrew, Thomas Joseph. *Memoirs of the Life and Writings of John Coakley Lettsom.* 3 vols. London: Nichols, Son and Bentley, 1817.

Phibbs, John. *Place-Making: The Art of Capability Brown.* Swindon, UK: Historic England, 2017.

Poliquin, Rachel. *Beaver.* London: Reaktion, 2015.

Poole, Steve. "Ghosts in the Garden: Locative Gameplay and Historical Interpretation from Below." *International Journal of Heritage Studies* 24, no. 3 (2017): 1–15.

Rembert, David H. "William Pitcairn MD (1712–1791): A Biographical Sketch." *Archives of Natural History* 12, no. 2 (1985): 219–29.

Rosner, Lisa. *Medical Education in the Age of Improvement: Edinburgh Students and Apprentices, 1760–1826.* Edinburgh: Edinburgh University Press, 1991.

Ruggles, Dede Fairchild, ed. *Sound and Scent in the Garden.* Washington, DC: Dumbarton Oaks, 2017.

Schiebinger, Londa. *Plants and Empire: Colonial Bioprospecting in the Atlantic World.* Cambridge, MA: Harvard University Press, 2004.

Schupbach, William. "Illustrations from the Wellcome Institute Library: Earl's Court House from John Hunter to Robert Gardiner Hill." *Medical History* 30, no. 3 (1986): 351–56.

Semple, Robert Hunter. *Memoirs of the Botanic Garden at Chelsea Belonging to the Society of Apothecaries of London.* London: Printed by Gilbert and Rivington, 1878.

Shapin, Steven. "Invisible Technicians." *American Scientist* 77, no. 6 (1989): 554–63.

Simo, Melanie. *Loudon and the Landscape: From Country Seat to Metropolis.* New Haven, CT, and London: Yale University Press, 1988.

Smit, Tim, and Anthony Kendle. "The Eden Project." *Acta Horticulture* 916 (2011): 153–58.

Spary, Emma. *Utopia's Garden: French Natural History from Old Regime to Revolution.* Chicago: University of Chicago Press, 2000.

Spooner, Sarah. *Regions and Designed Landscapes in Georgian England.* London: Routledge Research in Landscape and Environmental Design, 2015.

Stearn, W. T. "Sources of Information about Botanic Gardens and Herbaria." *Biological Journal of the Linnean Society* 3 (1971): 225–33.

Stewart, Larry. "Other Centres of Calculation, or, Where the Royal Society Didn't Count: Commerce, Coffee-Houses and Natural Philosophy in Early Modern London." *British Journal for the History of Science* 32, no. 2 (1999): 133–53.

Stobart, Jon. "'So Agreeable and Suitable a Place': The Character, Use and Provisioning of a Late Eighteenth-Century Suburban Villa." *Journal for Eighteenth-Century Studies* 39 (2016): 89–102.

Sweet, Rosemary. *Antiquaries: The Discovery of the Past in Eighteenth-Century Britain.* London: Hambledon and London, 2004.

Symes, Michael. *The Picturesque and the Later Georgian Garden.* Bristol, UK: Redcliffe Press, 2012.

Tarlow, Sarah. *The Archaeology of Improvement in Britain, 1750–1850.* Cambridge: Cambridge University Press, 2007.

Tinniswood, Adrian. *The Polite Tourist: A History of Country House Visiting.* London: National Trust, 1998.

Tobin, Beth Fowkes. *Colonizing Nature: The Tropics in British Arts and Letters, 1760–1820.* Philadelphia: University of Pennsylvania Press, 2005.

Wilkinson, K. D. "William Withering (1741–1799) and Edgbaston Hall." *British Heart Journal* 2, no. 4 (1940): 298–301.

Williams, Glyn. *Naturalists at Sea: Scientific Travellers from Dampier to Darwin.* New Haven, CT: Yale University Press, 2013.

Williamson, Tom. *Polite Landscapes: Gardens and Society in Eighteenth-Century England.* Baltimore, MD: Johns Hopkins University Press, 1995.

———. *The Transformation of Rural England: Farming and the Landscape, 1700–1870.* Exeter, UK: University of Exeter Press, 2002.

Wills, Kathy, and Carolyn Fry. *Plants: From Roots to Riches.* London: John Murray, 2014.

Withers, Charles. "William Cullen's Agricultural Lectures and Writings and the Development of Agricultural Science in Eighteenth-Century Scotland." *Agricultural History Review* 37, no. 2 (1998): 144–56.

Wulf, Andrea. *The Brother Gardeners: Botany, Empire and the Birth of an Obsession.* London: Windmill Books, 2008.

Wylie, John. *Landscape.* Abingdon, UK: Routledge, 2007.

Van Sant, Ann Jesse. *Eighteenth-Century Sensibility and the Novel: The Sense in a Social Context.* Cambridge: Cambridge University Press, 1993.

Vila, Anne, ed. *A Cultural History of the Senses in the Age of Enlightenment.* London: Bloomsbury, 2014.

Index

Note: Page numbers in italics refer to illustrations.